PACKY & ME

The Incredible Tale of Doc Maberry
and the Baby Elephant Who Made History

by Dr. Matthew Maberry
with his wife, Patricia Maberry
as told to Michelle Trappen

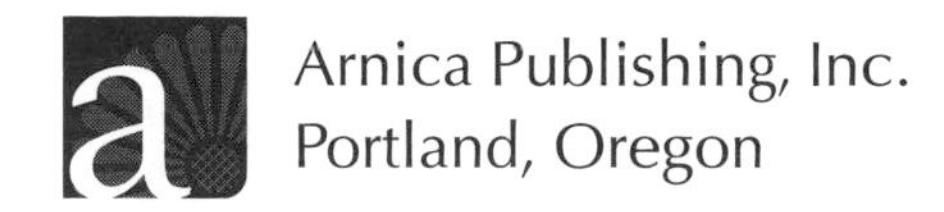
Arnica Publishing, Inc.
Portland, Oregon

Library of Congress Cataloging-in-Publication Data
Maberry, Matthew.
Packy & me : the incredible tale of Doc Maberry and the elephant who made history / by Matthew Maberry, with his wife, Patricia Maberry (as told to Michelle Trappen).
p. cm.
ISBN 978-0-9826401-3-5 (pbk.)
1. Packy (Elephant) 2. Asiatic elephant—Oregon—Portland. 3. Maberry, Matthew. 4. Veterinarians—Oregon—Portland. 5. Oregon Zoo (Portland, Or.) I. Maberry, Patricia. II. Trappen, Michelle. III. Title. IV. Title: Packy and me.
SF408.6.E44M33 2011
636.967092'9—dc22
2010051497

Arnica Publishing
3880 SE Eighth Ave., Suite 110
Portland, Oregon 97202

CEO/Founder: Ross Hawkins
Board Chair: Diane Vines
Production Director: Dick Owsiany
Design: Bo Björn Johnson
Developmental Editor: Jennifer Weaver-Neist
Managing Editor: Kathy Howard

Front cover photo: *In a rare quiet visit, Packy and I share a moment together. The baby elephant wrapped his trunk around my arm in a sign of affection that was very mutual. (Reprinted with permission from* The Oregonian*)*

To my beloved wife, Patricia

CONTENTS

ACKNOWLEDGMENTS

We gratefully acknowledge and appreciate the contribution *The Oregonian* newspaper has made for allowing the use of historic photographs and newspaper articles about Packy and other zoo-related activities, originally published in the *Oregon Journal* and *The Oregonian*. We also want to thank Mike Keele, director of elephant habitats at the Oregon Zoo, for his longtime support and for writing the foreword to this book; the late Shana Alexander for her firsthand observations reported in *LIFE* magazine and in her book *The Astonishing Elephant;* good friends Jim and Sue Anderson for use of their photographs and for Jim's kindness in writing the special "Note" for this book; and Marianne Marks, daughter of the late Jack Marks, the zoo director at the time of Packy's birth. In addition, we offer a special thank you to our proofreaders, Cathie Moravec, Jordan Mulholland, Kendall Mulholland, Candace Harris, and Patricia Schwock. Of course, much thanks to Arnica Publisher Ross Hawkins for understanding the vision of this project, and making the book a reality. And most of all, thanks to journalist Michelle Trappen for her hard work in helping us to write the story of Doc's life.

“[The elephant is] the beast which passeth all others in wit and mind.
—Aristotle

A mural of Packy graced the side of the Skidmore Fountain Building in Portland's Old Town from 1990 to 2008. The iconic painting, viewed from the Burnside Bridge, ultimately came down because of a building renovation. (Reprinted with permission from The Oregonian*)*

FOREWORD

By Mike Keele

I first met Dr. Matthew Maberry in 1971, when I began my zoo career at the Oregon Zoo, which was known then as the Portland Zoological Gardens. Under his supervision, I worked as an animal keeper providing care for animals being treated in the hospital and quarantine areas. I had very little experience, and Doc, who was patient and generous with his time and expertise, taught me a lot.

When Doc first learned that Packy's mother, Belle, was expecting, little was known about fetal development, neonatal care, maternal behavior, and a host of other details associated with captive elephant birth. I don't know if Doc appreciated—or knew—the impact his work would have decades later. If Packy were born today under the same circumstances, Doc's fame would be much greater. In the 1960s, however, zoos did not communicate well with one another, so a lot of information and breakthroughs—like Packy's birth—weren't easily shared. When I started at the zoo, it was still the only place in America where a captive elephant had been born. By then, though, Packy's birth had lost much of its initial luster. Once networking between zoos started in the 1970s, accomplishments like Dr. Maberry's allowed for greater personal recognition.

Many expert resources exist today thanks to the initial spark of Doc's work with elephants. The Association of Zoos and Aquariums

Chendra, among seven elephants currently residing at the Oregon Zoo, raised her trunk in gigantic greeting when I toured the facilities with Mike Keele during this 2007 visit. Cameras from the television show Oregon Field Guide *taped the day's event for a segment that aired on Oregon Public Broadcasting. Chendra's head looks huge because of the camera angle. (Photo by Sue Anderson)*

(www.aza.org) has a Species Survival Plan committee for elephants that coordinates breeding strategies, develops husbandry guidelines, and supports elephant programs with scientific panels, such as the Veterinary Scientific Advisory Group and the Reproduction Scientific Advisory Group. The World Association of Zoos and Aquariums' Global Elephant Management Program (www.waza.org) oversees captive elephant management worldwide, and the Asian Elephant Specialist Group (www.asesg.org) develops conservation strategies for elephants in range countries. These organizations, which are comprised of elephant experts, serve as incredible resources for all things elephant related.

Doc and I only worked together for about three years, and I didn't get the opportunity to work with the zoo's elephants until 1975, after Doc retired. But Doc's work has never been forgotten. Recently, in fact, the zoo reconnected with Doc after the discovery of home movies that had been kept in storage for many years. Because of my long tenure with the zoo and my interest in elephants, the movies were sent to me. What a treasure! The most amazing movie was of Packy's birth. Other clips showed moments leading up to the birth. I couldn't identify some of the folks in these movies, so I asked Doc to take a look to see if he could identify the unknown people and events. Fortunately he agreed.

It was a remarkable meeting! Beyond explaining medical techniques he attempted in 1962, Doc identified members of the medical community who were recruited to help with this unique pregnancy. He also described how Belle delivered and cared for Packy. We were so fortunate to capture his historic recollections on videotape.

Doc is a true pioneer in animal medicine, and even though it's been nearly fifty years since Packy's birth, he still deserves to be recognized for being the first American veterinarian—ever—to facilitate the birth of a captive elephant. I continue to marvel at the courage he exercised in allowing elephants to be elephants, trusting that they knew how to work as a herd to bring a new life into their world. It is truly amazing what he did. He blazed a trail for zoos and elephant conservation nationwide and around the world.

Mike Keele is the current director of elephant habitats at the Oregon Zoo and is recognized internationally for his Asian elephant expertise.

A NOTE ON MY FRIEND DR. MABERRY

By James O. Anderson

Dr. Maberry has spent his entire professional career caring for a variety of animals that represent the full spectrum of life, including people, encouraging the latter to have a better sense of appreciation for life. As a veterinarian, he is capable of using his human talents to help animals—from mice to monkeys, to elephants, eagles, and whales.

Dr. Maberry and I worked on several scientific, educational, and social projects together, all the way from releasing rehabilitated indigenous wildlife back to the wild, to educational programs for teachers and young people in schools and in the zoo, to creating educational programs and animal exhibits for the Children's Zoo in what was known in the '60s as the Portland Zoological Gardens. We searched for what were thought to be escaped snakes and other zoo animals in neighborhoods near the zoo, and at the same time educated our neighbors as to the values of indigenous and exotic wildlife.

In the winter of 1965, we flew around the state in his personal aircraft, investigating one of the largest southern movements ever recorded of the snowy owl (*Nyctea scandiaca*). We took part in pioneering work on the beluga whale (*Delphinapterus leucas*) in captivity. We designed and conducted educational programs for zoo personnel and teachers from all grades.

Good friend Jim Anderson and I visited in my home at one of my birthday celebrations. (Photo by Patricia Maberry)

As a man who believes in doing his best for the animal and human communities, Dr. Maberry has gone out of his way to educate everyone as to the complexities of animal and human ecosystems. It is in these areas that I find strength, and I continue to use many of these concepts in my work as a writer, teacher, and naturalist today.

To say that Dr. Maberry enriched the lives of thousands of people with his wise, forward-looking counsel and treatment of the animal community would be an understatement. Matt respects life in all forms and goes out of his way to ensure that humans and the so-called "lower animals" get the most out of it. It's also an understatement to say that I love him for the human being he is. When he departs from this Great Old Earth, he will leave a gaping hole in the lives of many.

James O. "Jim" Anderson is a naturalist, author, educator, and longtime friend of Dr. Maberry.

INTRODUCTION

By Patricia Maberry

When Matthew Maberry proposed to me in 1975, he promised that if I married him, I would never be bored. He has kept his word: our marriage has been anything but dull! Rather, my life with Matthew has been a whirlwind of activities and challenges—an adventure I never want to end.

Matthew is the most fascinating, intelligent, and knowledgeable man I've ever known; his always-curious mind never stops learning. Lucky me, I've been at his side the last three-plus decades, seeing and experiencing things that continually amaze me.

Animals, of course, have dominated our time together. But so did Matthew's inner circle—smart, driven people I lovingly call the "cast of characters." They include Jack Marks, the zoo director responsible for the construction of today's Oregon Zoo and one of the key players during Packy's birth; Morgan Berry, a multitalented animal importer who originally owned Packy's parents, Thonglaw and Belle; Eloise Berchtold, Morgan's business partner who had a national reputation for her exotic animal circus act; and Dr. Marlowe "Ditty" Dittebrandt, a Portland physician who segued to animal care, first working with Matthew at the zoo, then with Morgan at his home and private wildlife refuge, Elephant Mountain. None of these people lived standard nine-to-five lives; such hours skimped on their ambitions. I constantly marveled at their energy levels, their accomplishments, and—like Matthew—their love of animals. At Matthew's side I often worked with this cast of characters, but I knew I would never be one of them. I just felt happy, privileged, and sometimes a little intimidated being in their company.

Along with the cast of characters, Matthew also kept a "circle of friends"—professionals who revved his mental motor. Among them was naturalist Jim Anderson, who worked with Matthew at the zoo and joined him on many animal adventures; writer Shana Alexander, with *LIFE* magazine, whose article cast a national spotlight on Matthew after Packy's birth; trade show promoter Bob O'Loughlin, who formed a business with Matthew that focused on finding animals worldwide for display or research; Oregon Zoo's Mike Keele, who's now among the world's leading Asian elephant experts and who's always a gracious host during Matthew's annual zoo visits; and former Wilsonville resident Howie Renner, whose cows sometimes caused veterinary havoc but who has become one of Matthew's dearest friends.

When I attended Oregon State in the 1960s, I believed that college gifted a person with the bulk of their life's education, but life with Matthew disproved that thinking. The years with my husband have brimmed with constant education, much of it beyond my comprehension. Now, as always, I marvel at his memory and his ability to tackle life. He's built a cedar-plank boat, piloted his own airplane, and given mouth-to-blowhole resuscitation to a captured killer whale being flown from Alaskan waters to a Washington state aquarium. He's also written countless professional papers, several of which were published, detailing his work with animals or innovative techniques he's pioneered.

Now, a confession: when I first met Matthew, I thought he was rather rude. Based on a recommendation (actually it was a commandment from the breeder), I took my new poodle puppy to see a Dr. Matthew Maberry. This would have been in 1971, when Matthew still worked at the zoo. As usual, he had his private practice and treated most of his patients at night in the owners' homes. But daytime appointments he saw at the zoo's animal hospital.

I knocked on the animal hospital door and Matthew came dashing outside, his white lab coat flapping and the sun reflecting on the curl half-circling his forehead; he looked rather in disarray. Well, he gave my puppy a shot, asked me a couple of questions, gave me a couple of instructions, and told me to come back in three weeks. Then he closed the door.

I stood there aghast at this very abrupt man!

At my next appointment, though, he introduced himself and chatted. My little girl—Gabby—loved him. She sat in his lap and kept licking him.

The third time I took Gabby back, Matthew asked me to volunteer in the zoo's clinic. Anybody who knew me then would have laughed at the thought! I drove a hot little sports car, and wore suits and heels to my buyer's job at Meier & Frank, a former chain of department stores in the Pacific Northwest. I was far from the perfect candidate to clean up animal waste.

But I said yes. And it was marvelous, especially when I got to work with Matthew. That didn't happen for a few months. I think he asked me—no, *told* me!—to get my white lab coat because we were headed to the elephant barn. I had to trot to keep up with him. Once there, we performed vaginal smears on the elephants, and it was my job to hold the glass slides while he did the back-end work.

Another time when Matthew and I worked in the elephant barn, two of the elephants needed to know for sure if I was a boy or a girl. So they stuck their trunks up my skirt into my panties and down my shirt

In 1975 Patricia and I visited the Taj Mahal on our honeymoon, which also doubled as a business trip. (Courtesy of Dr. Matthew and Patricia Maberry)

into my bra! I tried not to scream or squirm. Matthew, watching this, told me not to worry, that they wouldn't hurt me. And, of course, they didn't.

Some of the funniest times were the days I went to work at Meier & Frank—after volunteering—and people told me, "You don't smell very good." I'd say, "I showered this morning, but I've been up in the elephant barn." (I didn't have to explain any more than that.)

I have frequently told Matthew and the cast of characters that each was worthy of a book; I knew their larger-than-life personas would entrance readers. But, of course, all were way too busy living life to tackle a book.

Through the years, several people approached us about writing Matthew's book, but the time never seemed right. Now, as I share my husband's twilight years, I realize time is running out. Nine years ago I began the daunting task of writing the legacy of his life and that of his friends but never finished. Finally, in 2010, we found the right people to make this book a reality.

I didn't know Matthew in 1962, when he delivered Packy and made animal history; I wish I had. My husband has experienced many highs in his long life, but Packy's birth tops everything.

Matthew didn't seek fame; it found him. It's sort of how I discovered him—coincidentally. He wasn't looking for Packy; I wasn't looking for him. But in both situations, we listened to our heads and our hearts.

And let nature lead the way.

Her sides bulging with baby weight, Belle glared at cameras attempting to capture her maternal glow. Belle kept me and the rest of the world guessing her baby's birth date. (Reprinted with permission from The Oregonian*)*

The Great Portland Elephant Watch

In the frosty, dawn hours of April 14, 1962, a hairy, pink baby elephant plopped back-feet-first onto a straw-covered floor at the Oregon Zoo, then known as the Portland Zoological Gardens.

Not one single press camera clicked—surprising since the media lived in the elephant barn around the clock for more than three months just to capture this moment. Magically, though, a zoo movie camera quietly recorded Packy's historic birth. I say "magically" because I was so focused on ensuring a safe delivery that I never noticed the filming. Truthfully, I was glad the press wasn't there, causing noise or popping flashbulbs around Belle, the mother; I didn't want any interruptions that might jeopardize this delivery. But I'm sure glad now that at least

On the job at the zoo, I often wore coveralls, but after work, at social functions, I always dressed nicely. (Courtesy of Dr. Matthew and Patricia Maberry)

one camera rolled so that the world, like me, can still enjoy seeing Packy's first wobbly moments of life.

My name is Dr. Matthew Maberry, though most folks call me Doc. In 1962, as the zoo's first-ever full-time veterinarian, I delivered Packy, the first captive elephant born in an American zoo.

To say that this birth was a big deal is an understatement of elephantine proportions. Today, baby elephant photos regularly pepper newspapers and television screens, but in 1962, such photos didn't exist. *Baby elephants didn't exist*, at least not in Western Hemisphere zoos or circuses. Hormonal male elephants proved too cantankerous to keep,

To see footage of Packy's birth, visit this link at the zoo's Web site: http://www.oregonzoo.org/VideoArchive/PackysBirthday.htm

and females couldn't make babies all by themselves! Captive-birth practices were far from ideal as well. Before Packy was born, the handful of elephants that were born all died in infancy or shortly thereafter.

Historically, it was a time of unease and upset in the United States, with the Vietnam War decreasing in popularity by the day, racial tensions erupting across the country, and Cuba knocking at our southern border with the Cold War. We needed some joyful distraction.

Packy's birth delivered 225 pounds of happy news and established Portland's zoo as the world's preeminent captive breeding facility. In coming decades, scores of baby elephants would be born in zoos and circuses throughout the United States and beyond, each a success to be celebrated. In turn, I became a renowned authority on elephant birth, and Packy became Oregon's beloved mascot.

Today, captive elephant breeding has become controversial, and with good reason. Some people think that zoos and circuses exploit exotic animals, and deprive them of a normal life. Believe me I see the logic in that argument. All animals, humans included, prefer their natural habitat.

In 1962, however, few people thought that way. Exotic animals were something fresh and exciting—a far cry from the ho-hum barnyard scene. Nobody, scientists and veterinarians included, knew much about wildlife from faraway lands like Asia, Africa, and Antarctica. The general consensus was that if these animals—like dogs and cats—had enough food, a safe place to sleep, and basic medical care, their needs were being met.

Exotic animals thrilled and terrified us. We wanted to see them up close, but not too close. Elephants, the world's largest beasts, especially mesmerized the public—and equally frightened their keepers. I can say from experience that an elephant in labor is a somewhat scary sight, given the rocking and spinning and crying that goes on. I'm sure that's why keepers who assisted with this country's earliest elephant births chained the mothers—out of pure fear.

That's another reason why Packy's birth struck such a chord. A baby elephant was born and survived because we figured out how to do it the way nature intended: unencumbered and in its due time. It was a breakthrough in veterinary medicine and zoo philosophy alike.

In the months leading up to Packy's birth, I wouldn't say I was nervous; worried is a better word. Determined might be the best word. Deeply, I felt the importance of this birth, both for Oregon and veterinary science. Nothing this headline-grabbing had happened since the Lewis and Clark Centennial Exposition in 1905, which in following years boosted the state's population by more than 100,000. I knew that if Belle delivered a healthy baby without complications, and without being chained, this captive birth would teach what couldn't be taught before.

Since my own birth in 1917, I've been a problem solver—the type who takes stuff apart and figures it out. My wife, Patricia, will tell you about the many times she's come home to find our Chrysler's motor dismantled and lying on our garage floor. Just fixing something doesn't work for me; I need to know the source of the problem, its evolution. Then, for me, life makes sense.

I suppose that's why Packy's birth became so important, both to me and to the rest of the Western Hemisphere. We were learning something new—something *big*—in the animal world; it wasn't just a birth. It was a learning expedition, a collective caring, a step forward in veterinary and zoo sciences.

Lucky me, I had a front row seat to what has gone down in history as the Great Portland Elephant Watch.

Elephant Breeding 101

Now before I share the particulars of the Great Portland Elephant Watch, first let me explain what I knew about elephant breeding and birth before Packy's arrival:

Pretty much nothing.

Of course I did my homework; I've always been a sharp student. I was born in Seattle, but in the mid-1920s my folks moved our family to a 160-acre dairy farm in Sequim (pronounced "SQUIM"), Washington, which then was a blip of a town skirting the north end of the Olympic Peninsula. Then, like now, I loved learning. Typically I kept a book tucked in my pocket so I could read in between cleaning stalls or milking cows. At night, because our two-story log house lacked electricity, I devoured books by kerosene lamplight. A quick study with a photographic memory, I could remember a lecture, verbatim, for a week. I finished enough coursework to graduate from Sequim High School at age twelve. But my mother, who worried that I was not mature enough to move on, insisted I stay in school and graduate with the rest of my class in 1935.

My family fared better than most families during the Depression. When Wall Street crashed in 1929, my father sold half our land (eighty acres) at a minimal price to cash-strapped neighbors and destitute friends. Our dairy business went on as usual; folks still needed milk, butter, and cream, even in hard times. I never went hungry. We always had lots of pigs, rabbits, chickens, turkeys, and geese, most of which we slaughtered and hung in the smokehouse. My mother also planted and nurtured a huge garden. Anytime a neighbor or relative visited they always left with a couple of hams and an armload of vegetables.

Every family member pitched in during the lean years—including me. Instead of heading to college like I wanted I worked on a neighbor's dairy farm, earning $90 a month. Nearly every dime went to my parents. Finally in 1937, at age twenty, I applied to medical school at Washington State University. I always wanted to be a doctor; my family descends from a long line of physicians. But when the medical school rejected my application because I could not pay the first year's tuition up front, a family friend steered me toward the university's veterinary school, where tuition could be paid over time. It took me a decade to finish—family issues and World War II slowed my progress—but in 1947 I finally graduated from Washington State among the top students in my veterinary medicine class.

But none of my farm upbringing or years of veterinary training included any instruction in elephant breeding and birth. Elephants were something folks saw in circuses or maybe at the zoo. And elephant keepers in those facilities basically winged it when a baby was born—typically with disastrous results. As writer Shana Alexander shares in her book *The Astonishing Elephant*: "The little that was known of the abilities and habits of elephants was a ragbag of lore and anecdotes handed down by unlettered circus hands, trainers, zookeepers, *mahouts* [handlers in India], *oozies* [handlers in Burma], ranchers, historians, and a few fabulists."

The one thing I did know was that Belle was definitely pregnant. In December 1961, Morgan Berry, an animal importer who lived near Seattle and owned Belle, felt fairly certain she was expecting; he had witnessed Belle and his bull elephant, Thonglaw (pronounced "TONGUE-la"), mating in July 1960. But Morgan was not 100 percent sure. Pregnancy tests for elephants did not exist, and no American vet—me included—had ever made such a diagnosis. Unlike pregnant women, whose bellies bulge after a few short months, an extra 1,000 pounds on a 6,000-pound elephant barely shows. Only in the final months can the human eye detect growth, as well as fetal kicking. It took vaginal smears and a pelvic examination of Belle's blood vessels for me to confirm that a baby was growing within her.

That's when I became like Sherlock Holmes, searching everywhere for clues about captive elephant breeding. It was a frustrating hunt. Little had been published about elephant birth, and the Internet—nowadays bursting with facts—did not exist. I found one book, *Patterns of Mammalian Reproduction*, by S. A. Asdell of Cornell University, but it dedicated just a page and a half to elephants, much of it useless. Asdell wrote that gestation lasted anywhere from 500 to 720 days, meaning between 17 and 24 months—a huge gap of time for a vet to be at the ready. Another book, *The Care and Management of Elephants in Burma*, by A. J. Ferrier, informed: "From the twelfth to the sixteenth month, a pregnant female can do light work but should on no account be made to *aung* in streams during rises, as she may be

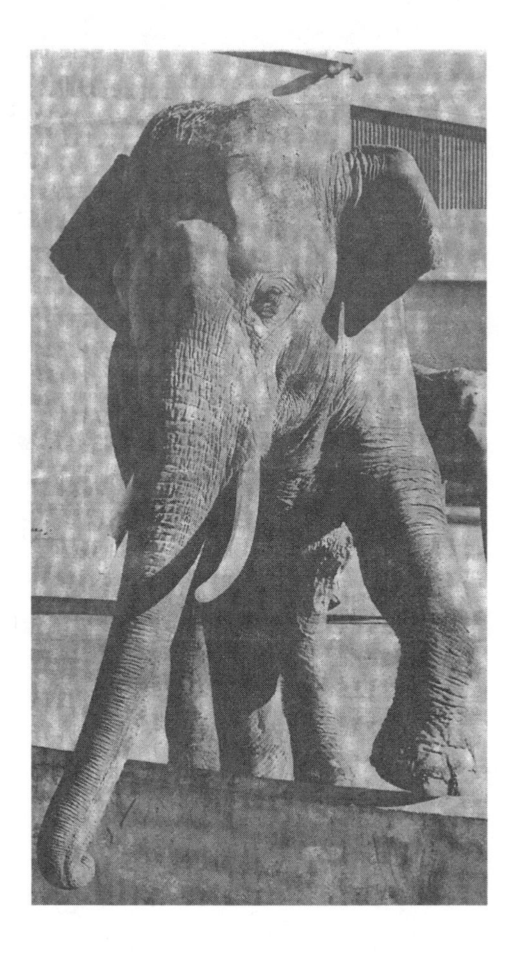

The final few months of Belle's pregnancy, Thonglaw often looked as he does in this photograph: mean, menacing, and ready to charge anyone who came near. Thonglaw suffered from an annual hormonal affliction known as musth, *which causes deep, unexplained inner fury. (Reprinted with permission from* The Oregonian*)*

struck by floating logs." I wasn't sure what "aung" meant, but I didn't think Belle would be doing it at Portland's zoo.

I did learn about the first elephant to ever set foot on United States soil. The unnamed elephant arrived in April 1796 on the trading ship *America*. Captain Jacob Crowninshield, who bought the young pachyderm for $450 in Bengal (modern day Bangladesh), soon sold his purchase—for $10,000—to a John Owen of Charleston, South Carolina. Mr. Owen obviously recognized colonial America's thirst for entertainment, because he advertised

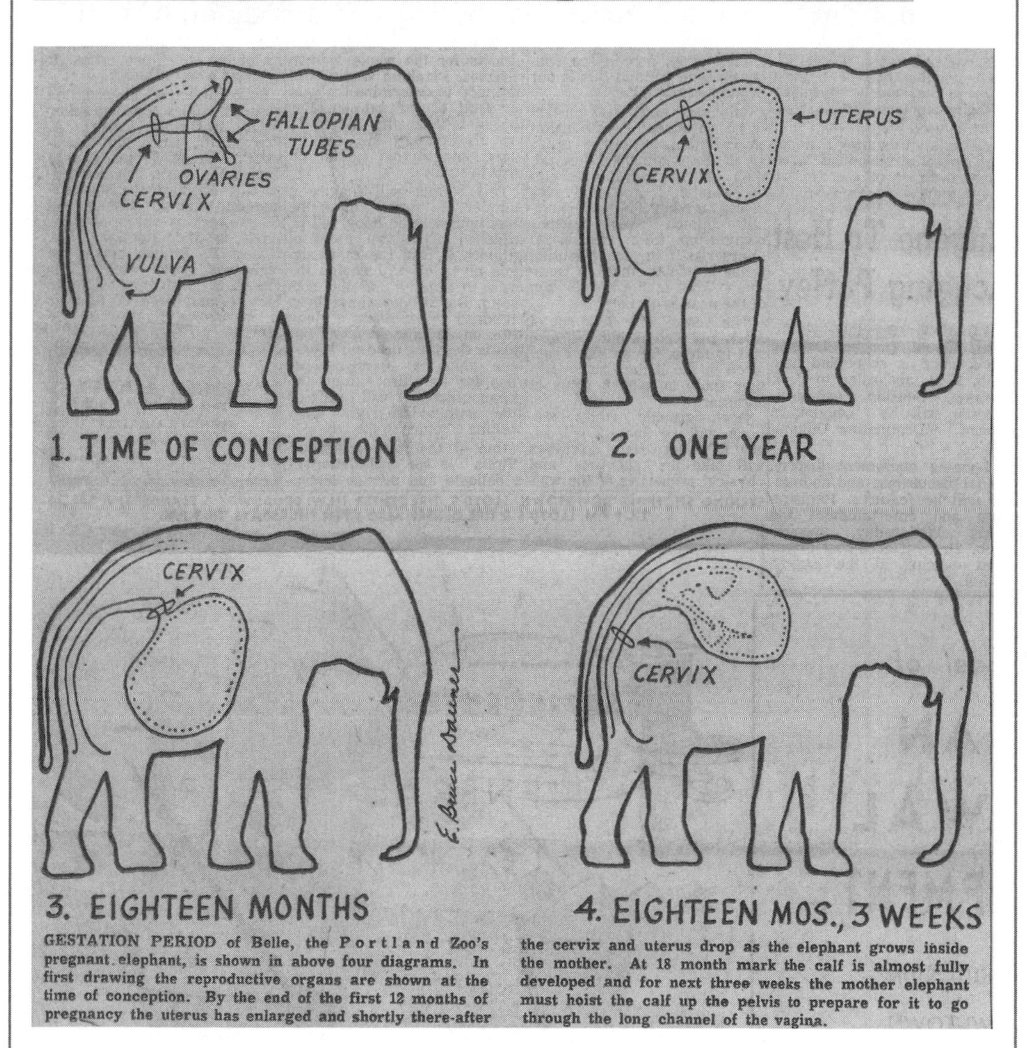

Veterinarian Explains Events Leading To Elephant Birth

GESTATION PERIOD of Belle, the Portland Zoo's pregnant elephant, is shown in above four diagrams. In first drawing the reproductive organs are shown at the time of conception. By the end of the first 12 months of pregnancy the uterus has enlarged and shortly there-after the cervix and uterus drop as the elephant grows inside the mother. At 18 month mark the calf is almost fully developed and for next three weeks the mother elephant must hoist the calf up the pelvis to prepare for it to go through the long channel of the vagina.

In February 1962, The Oregonian *used this graphic to explain elephant gestation to readers. (Reprinted with permission from* The Oregonian*)* in newspapers everywhere: "Come to see The Elephant! 25 cents!" And people did—in hordes.

My research also taught me that just six captive elephants had been born in the United States—*ever!* The first arrived on March 10, 1880, to parents Hebe and Mandarin, who mated in London while working for The Great London Circus. As luck would have it, The Great London Circus became part of Cooper, Bailey, and Co.'s Circus in the United States before the birth, and when Hebe (better known as Babe) delivered Little Columbia, nobody could have been happier than circus co-owner James A. Bailey. Rival circus owner P. T. Barnum reportedly offered $100,000 for the newborn; Bailey wired back, "Will not sell at any price!"

The second captive elephant baby, born on February 2, 1882, scored a lengthy story in *The New York Times*. Beneath the huge headline "BARNUM'S BABY ELEPHANT," the reporter meticulously detailed the birth of the smallish, forty-five-pound baby, who was named Bridgeport after her birthplace in Connecticut. The reporter's story exemplifies how elephants were historically handled during birth.

> *At eight o'clock last evening, twenty elephants of various sizes were chained by one of their hind legs to stout posts in a circle close to a sawdust ring. Eight minutes later, Queen, who was chained to a heavy post in the center of the ring, was delivered of an heir, whose entree into public life was greeted with a succession of roars from all the elephants in the building. Drs. Cole, Sandford, Porter, Bronson, and Hubbard, of this city, soon arrived, and everybody, excepting the elephants, was disposed to merrymaking. The elephants seemed to take a gloomy view of the situation and, while continually swaying to and fro, winked in an ominous manner or flapped their blanket-like ears, as though preparing to strike terror to the hearts of those of the human race who stood close to the outer circle of the ring.*

The third, fourth, fifth, and sixth captive babies all belonged to the misfortunate Princess Alice (so named for President Theodore Roosevelt's daughter Alice) and Snyder, a captive elephant couple with the Sells Floto Circus based in Denver, Colorado. In 1910 the pair courted and bred to the delight of circus owners, who knew well enough the potential profits of a baby elephant on tour. But in 1912, when Alice went into labor, elephant handlers separated her from the other elephants and staked all four of her legs to the ground. Afraid and alone, Alice delivered, but never having witnessed another birth, the startled mother tried to kill Baby Hutch, as the newborn male was named.

Circus officials quickly saved the baby and then tried raising him on a concoction of boiled rice, cow's milk, and condensed milk. For six weeks the baby lived, and attracted throngs of spectators. But one night a fire engulfed the circus train, and Baby Hutch, badly bruised, soon died. An autopsy showed that the rice diet was clogging Baby Hutch's digestive system, and had he survived his wounds, the diet would have killed him sooner rather than later.

"Angry" Alice, as she became known, and Snyder successfully mated three more times, but circus handlers always chained and isolated the terrified mother during labor. All three babies died at birth or soon after. The final infant, Prince Utah, was born in 1918 and lived nearly a year, until his mother accidentally rolled onto him and killed him. Newspapers reported that Alice cried at the discovery. Whatever happened, Prince Utah's birth became a significant marker in captive elephant birth history: forty-four years would pass before the Western Hemisphere witnessed another birth—Packy's in 1962.

Circus and zoo owners might have known more about elephant birth if they owned male elephants. At the time of Packy's birth, just thirteen male elephants existed in the United States. The reason: males periodically enter a hormonal phase called "musth"

In 1918 a Roseburg, Oregon, resident photographed Princess Alice, a Sells Floto Circus elephant and her fourth baby, Prince Utah, in Salt Lake City. Prince Utah outlived his three older siblings, who all died shortly after birth, but not by much; he passed away before his first birthday. Prince Utah was the last elephant born in captivity until Packy. (Reprinted with permission from The Oregonian*)*

("MUST-tha") that makes them a little crazy for anywhere from a few weeks to several months. *Musth*, originally a Persian term, means "intoxicated"; in Hindi, it means "madness." A male elephant in musth can get so worked up it will kill, including humans. In 1898, for instance, a raging bull elephant named Prince attacked and killed a circus show's beloved elephant boss. In retaliation, distraught elephant keepers slowly and cruelly destroyed Prince by stabbing him with pitchforks.

Today, zoos like the Oregon Zoo have modern facilities that allow the segregation of bull elephants while they're in musth. In earlier years, however, most zoos and circuses had no such facilities, so they typically showcased only docile female elephants, most of which were captured in the wild and purchased overseas. Consequently, little breeding went on. The few captive male elephants that were in the United States were heavily chained or led monastic lives. "Not a single one could ever even get near a female," Morgan Berry told *True* magazine in 1969, "and if they could have, the poor devils wouldn't have had the get-up-and-go to do anything about it—they were so poorly fed and cared for."

In the world's Eastern Hemisphere, however, circus and zoo officials must have figured out how to handle occasionally grumpy male elephants, because between 1918 and 1962, forty-five captive baby elephants were born. European and Asian countries, all with much longer histories than young America, were located closer to Africa and southern Asia, native habitats of elephants. These births, though, had their share of troubles; seven babies were stillborn, six died before the age of one, and three were killed by their mothers shortly after delivery. But many elephants survived and went on to perform in circuses or be viewed in zoos; one elephant born in 1925 lived to be seventy-three years old.

Getting information on those births proved nearly impossible in 1962. I'm not sure that an Asian or a European elephant studbook, which keeps track of elephant births, even existed; the *North America Regional Studbook—Asian Elephant* was not published until 1985. (Mike Keele, now director of elephant habitats at the Oregon Zoo, published that book and has since continued in his role as studkeeper.) Also, most of the Eastern Hemisphere births were in communist countries; I highly doubt circus or zoo officials there would have shared any elephant knowledge. Such was the world's political climate at the time.

Fortunately, I discovered that several captive baby elephants were born at Rome Zoo between 1948 and 1955. Armed with an interpreter, I called the zoo's Dr. Ermanno Bronzini, who I understood had helped deliver at least one of those babies. Portland's phone company, enamored with the upcoming birth (like the rest of Oregon), offered to pay for unlimited conference phone calls to Italy. Dr. Bronzini could not be reached, but I was referred to another Italian, Dr. Emmanuel Amoroso, a professor emeritus working at the Royal Veterinary College in London. Dr. Amoroso did not have a telephone, and it took awhile to track him down. Finally, with Portland's Wilson High School language teacher Leonora Guinazzo at my side, I spoke with Dr. Amoroso, who had witnessed the births of African elephants in Ethiopia in the 1930s. His best advice: watch for a sudden drop of 1 to 3 degrees in Belle's normal temperature of 98 degrees Fahrenheit; that meant birth was twelve to twenty-four hours away.

I appreciated Dr. Amoroso's knowledge of African elephant births, but I wasn't sure if it applied to captive Asian elephants. Just in case, I took Belle's temperature daily, a chore that might not appeal to everyone. Shedding my shirt and undershirt, I smeared Vaseline on my right arm up to my armpit and shoved a large-animal thermometer, lashed onto a long spearlike dowel, into Belle's rectum. Belle's temperature,

Yes, I felt as gloomy as I looked in 1962, when a phone call to Rome failed to reach a much-needed elephant birth expert. Wilson High School language teacher Leonora Guinazzo then translated a second phone call to London, where I successfully connected with Dr. Emmanuel Amoroso, another elephant expert, who also had few answers. (Reprinted with permission from The Oregonian*)*

though, remained pretty much normal in the long months leading up to delivery. And when she was in active labor, I wasn't about to interfere by sticking anything inside her!

Really, my best source of information about the *Elephas maximus* came from Morgan Berry. Before Morgan bought Belle he spent a great deal of time in Asia, gleaning any information he could find about the species. In India, in the city of Sonepur, Morgan attended an annual *mela*, a fair of ancient origins, which featured more than 500 elephants. There, he spent two weeks measuring and photographing elephants.

More importantly, since he planned to breed, he learned about lovemaking. Elephants, Morgan discovered, are a lot like humans in that they court and fall in love. Then they practice mating, with the male mounting the female but not actually penetrating. Finally, the pair consummate, at least once, and possibly three or four times over a twenty-four-hour period. The honeymoon phase lasts for months afterward.

The *mahouts* told Morgan that elephants definitely do not breed while they are chained. They typically journey far into the forest, away from human eyes, to court and mate.

The little that I had learned about elephant delivery led me to one definite conclusion: Belle should not be chained during labor. Like other herd animals, elephants naturally depend on one another for support and friendship. Chaining only frightens an elephant—especially a pregnant elephant. I also figured out that a mama elephant needs her best friends, or "aunties," during labor and after delivery.

My somewhat futile research attempts left me stumped but not defeated. I knew Belle would eventually solve this pachyderm puzzle . . .

By giving birth.

"If you've ever worked in a zoo, you never get over it. There is something about animals that really gets you in the heart.

—Doc Maberry

Morgan was a true Dr. Dolittle, the way animals followed him around. More so, Morgan was a man's man. He lived life to the fullest, eyes always focused on the next great adventure. I understood him well, since I walked the same road. (Reprinted with permission from The Oregonian*)*

Elephant Man Morgan Berry

As I've said, few folks knew much about elephants and how they gave birth in the early 1960s. But one man knew more than perhaps anyone in the United States. Lucky for me, Morgan Berry was my good friend.

Morgan and I got to know each other because he boarded three of his elephants at Portland's zoo in the winter. Summers, Belle, Thonglaw, and Pet lived at Seattle's Woodland Park Zoo, where they gave rides and performances. I had vet school training, but Morgan had firsthand knowledge of elephants—especially rare at a time when only 300 elephants lived in American zoos. Our friendship evolved because of a mutual love of animals and a respect for each other's opinions.

Morgan and I became closest during Belle's pregnancy. Elated, we were also concerned. Anything could happen; circus elephant births

"The first time I saw Morgan Berry in action, he was tucking a two-ton elephant into bed for the night. Belle was restless and apprehensive, beset by labor pains. The door popped open and in walked Morgan Berry, her devoted owner. Berry . . . strode through the bars and embraced the agitated elephant around the trunk. Soon he lay down on the concrete floor of the elephant house where he could "hold hands" with big Belle through the bars of her boudoir. She rumbled contentedly, lay down beside her soul mate, and fell sound asleep.

—Lev Richards, *The Oregonian*, 1962

and subsequent deaths proved that. Ultimately, Morgan and I decided to go with our gut feelings about animals and let the pregnancy evolve as naturally as possible, just like it would in the wild.

Morgan's thinking came from knowledge he gleaned on multiple trans-Pacific voyages to eastern Asia in the 1940s and 1950s. A short, balding man with Walter Cronkite glasses, Morgan no doubt shocked his wife when he up and decided to become an animal importer. Up to then, music dominated his life. Trained at the Juilliard School in New York City, he could bang, pluck, or blow almost any instrument, especially the trumpet. In the 1920s he played piano and organ in silent movie theaters, and in the 1930s he boogied as a jazz band drummer and trombonist, often aboard trans-Pacific cruise ships.

But in 1948, at age thirty-seven, Morgan took a job as the sole fireman on an aged lighthouse tender vessel headed for Hong Kong. Once there, he plunked down his life savings of $7,000 on 30,000 goldfish and an assortment of monkeys and tropical birds, all of which he planned to sell for a profit in the United States. The return trip, though, ended in disaster. The freighter stopped for seven days in Formosa (now known as Taiwan) to unload nitrate, which explodes if touched by water. No water was allowed to board the ship for a week—ghastly news for Morgan; his moneymaking goldfish required 2,000 gallons of fresh water a day! The fish, the fortune—all perished before Morgan got home to Seattle.

Such a setback would dissuade most from future endeavors, but Morgan was no ordinary man. Much to the dismay of his wife, at home with the couple's four children, Morgan mortgaged his house and again worked his way across the Pacific. This time, however, he already had a customer. Woodland Park Zoo wanted him to be their main supplier of exotic animals and even offered free, temporary cage space for larger animals that he planned to sell elsewhere.

"That's when I started thinking seriously about becoming an elephant breeder," Morgan told *True* magazine in 1969.

Morgan Berry's Elephant Legacy launched in March 1952 with the purchase of one-month-old Belle. "She was born in a jungle camp just outside Bangkok in February. I bought her in March, but it was June before I got home," Morgan says in the 1996 book *Elephants Don't Snore,* by writer Lev Richards.

Morgan had learned from India's *mahouts* that the bond between elephant and man can be extremely tight, lasting a lifetime since both live approximately the same number of years. Knowing this, Morgan raised 35-inch, 185-pound Belle as if she were his daughter.

Morgan and his family lived in a two-story house in the Fremont district of Seattle. "They had a big yard carefully screened by big trees, where Berry's ever-changing menagerie of wild animals, birds, and reptiles could indulge in a little outdoor exercise," Lev Richards writes in *Elephants Don't Snore*. "Nervous neighbors grew acclimated, if not accustomed, to the snarling of leopards, the screaming of cougars, the chatter of monkeys, and the roar of lions."

Belle, who was potty-trained, had free reign of the Berry house but never wrecked furniture. Like other young elephants, she craved attention and fretted when a family member was not in sight. When happy, she squeaked with shrill delight, especially when shadowing Morgan, his wife, or the children. "[She followed us] so close she would step on our heels and pull our shoes off," Morgan told *The Oregonian* in 1962.

Belle slept in the basement on a hay pile fenced in by cots. Morgan's sons, Kenneth and Howard, slept on the cots, as it was their job to wake up every two hours to feed the wood furnace. Some nights, though, the alarm clock would ring, waking Belle but not the boys. Wanting attention, Belle somehow climbed over the sleeping boys and snuck upstairs to the second floor, where Morgan and his wife slept. "Belle would tap us with her trunk until we fed her," Morgan told *The Oregonian*. "I would lie there and toss [treats] into her mouth until she quieted down and went back to bed."

Belle got the most attention when the family took her on evening walks around the neighborhood or when they piled the pachyderm into the back of Morgan's pink convertible Cadillac and cruised around town. Passersby openly gawked as Belle lovingly coiled her trunk around Morgan's shoulders. Once, Morgan's sons walked Belle to school for show-and-tell and nearly caused a car accident. "Cars came screeching to a halt, creating a traffic jam as motorists stared in disbelief," Morgan told Lev Richards.

At one year of age Belle stood five feet tall and weighed 660 pounds, finally making it difficult for her to maneuver the Berry household. That's when Morgan moved her to the nearby Woodland Park Zoo and ramped up his hunt for the perfect male elephant, Belle's future suitor. The *mahouts* had told him what to seek: a well-built male with an even-humped back; large, clear eyes; a trunk of good girth; and tusks of solid, white ivory. Powerful and potent, yet manageable.

Morgan found such an elephant in Thailand. "As soon as I set eyes

Seattle Trainer Says Elephants Need, Thrive On Tender, Loving Care

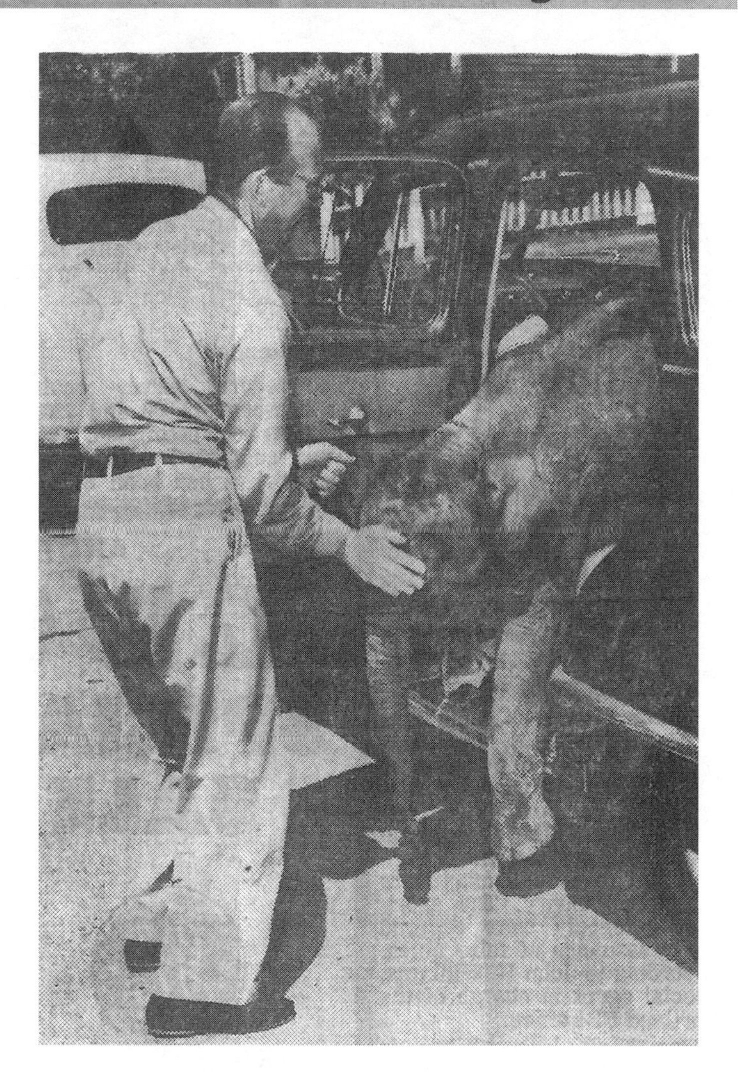

Morgan enjoyed telling people how he drove around Seattle with Belle in the backseat of his pink Cadillac convertible. What a character! In this picture, shot by The Seattle Times, *Packy's future mama is about six months old. (Reprinted with permission from* The Oregonian*)*

Zoo On Full Alert For Elephant Calf

In January 1962, when we told the media that Belle was expecting, we did so because we thought the birth was imminent. Here, Morgan escorted Belle around the "delivery room," hoping that the stroll would speed up what appeared to be immense labor pains. Belle's condition attracted a lot of empathy from the other elephants. (Reprinted with permission from The Oregonian*)*

on him, I knew that this was the one. I had to have him," Morgan told *True* magazine in 1969.

Thonglaw, then five, belonged to a Greek who, for some unexplained reason, had been kicked out of Cambodia with just his clothes, his young daughter, and his elephant. Shrewdly, Morgan swapped $20, two bicycles, and a baling machine for the "well-built" elephant.

Thonglaw's name came from Morgan's silent partner, Thonglaw Punyanitra, who was an attorney, money manipulator, munitions runner, and general exporter-importer. When Morgan told his backer of his plans to make his new elephant more famous than Barnum's Jumbo, Punyanitra humbly asked that the elephant be named for him.

Morgan chose wisely. Fertile Thonglaw went on to father an astounding fifteen surviving children, crowning him as the most valuable bull elephant ever in captivity.

Two years later, also in Thailand, Morgan bought baby elephant Pet, who grew up to be a nymphomaniac! Unlike the other female elephants, who courted Thonglaw only when they were in heat, Pet rubbed up against him every chance she got. I think Pet's constant need for sexual attention motivated Thonglaw's virility.

The key to captive breeding, though, was Morgan's success at finding a way to handle Thonglaw's periodic bouts of musth, which began at age ten (three years before Packy's conception). For about six weeks each year, Thonglaw had to be locked up alone in an isolation cell. As Shana Alexander wrote in her book *The Astonishing Elephant*, Thonglaw became a "homicidal maniac who might charge any man who approached him, even Berry, with intent to kill."

To bring Thonglaw out of musth, Morgan did what today sounds crude and somewhat cruel. In the infant years of exotic animal study, however, it worked most effectively. Morgan got a transformer from a neon sign and mounted it on the end of a long stick. He'd get in close to Thonglaw and zap him with that until the rattled animal was up against the wall. After that, Thonglaw quieted down because the prod really gives a shock. Before long, Thonglaw would be back to his old gentle self—until the next annual bout of musth.

Eventually, Morgan's animal inventory expanded to include dozens of species. I visited and cared for zoo animals daily, but Morgan literally lived with them. In 1965, wanting to be nearer to Portland's zoo, Morgan bought an eighty-acre, sprawling spread near Woodland, Washington, thirty minutes north of Portland. It was quite the sight visiting what came to be known as "Elephant Mountain." Elephants, horses, camels, hippos, zebras, and giraffes all roamed the forested acreage; rhinos, big cats, and bears lived safely in outside pens; and woodland mammals, snakes, and marsupials occupied cages. Elephants had their own concrete and steel barn, but babies always got their bottles in the Berry kitchen. The population frequently changed as Morgan bought and sold his inventory.

It was a huge job taking care of the 100 or so species that lived on the farm at any given time. Morgan, heavily assisted by his sons, cared for all of the animals, especially the elephants, who required food every four hours. Morgan's neighbor, Joe Wodaege, who watched Morgan feed the elephants many times, told writer Shana Alexander, "You gotta feed 'em when you get up, feed about noon, feed again about four, then about ten at night you give 'em an extra load of hay. That's nine loads of hay every four hours. Then there's the other end—you gotta shovel!"

Between 1975 and 1979, after my wife, Patricia, and I wed, we visited Morgan and his family at Elephant Mountain frequently, usually on Sunday afternoons. They lived in a Swiss-like chalet built by Morgan and his sons, with wall-size windows staring at the nearby Lewis River and the more distant, ribbonlike Columbia River. Morgan,

Eloise Berchtold amazed me. When I met her in the mid-1970s, she was known as the greatest female animal trainer in the United States. Then Morgan's business partner, she lived and trained on his property at Elephant Mountain. I'm good at healing animals; Eloise was good at teaching dangerous animals incredible tricks. I admired her talent and courage. (Courtesy of Dr. Matthew and Patricia Maberry)

a true renaissance man, always cooked some wonderful meal and then wowed us with a key-banging performance on his huge Hammond organ.

Morgan's life was never dull; like many ambitious men of our generation, he ran at full speed. Challenges needed conquerors.

Yet, sadly, life has its wounds. Patricia and I still feel sorrow recalling the night in May 1978 that our good friend Morgan needed help.

Raised in simpler times, Morgan paid for everything in cash or with in-store credit; he never carried credit cards or wrote checks. So when he needed a plane ticket fast and lacked cash because banks were closed, Morgan leaned on his closest friends. Of course, we raced to the airport and emptied our pockets.

It was, indeed, an extremely tragic day. Morgan's longtime business partner and very close friend, Eloise Berchtold, had been gored and killed by an elephant while performing for the Gatini Circus in Rock Forest, Quebec. Eloise grew up in Cincinnati, the daughter of parents who were passionate about wild animals, and a similar passion no doubt persuaded young Eloise to run away from home at age fifteen to perform with the Ringling Bros. and Barnum & Bailey Circus. Her considerable talents propelled her through the ranks to eventually land her a headline act showcasing bears, big cats, and elephants. By the 1970s, when Eloise lived and trained at Elephant Mountain, she had earned the widely printed reputation as "The World's Greatest Female Animal Trainer."

Several versions exist regarding Eloise's death. Personally, I think Eloise tripped, perhaps during a fast waltz act in the small circus ring. Teak, among Eloise's favorite elephants, bent to help, and accidentally gored his beloved trainer in front of a horrified audience. Loyal Teak then stood guard over Eloise and refused to respond to commands from circus officials and local authorities. I'm not sure what happened next, but Teak was either shot by authorities or euthanized by circus officials. No matter how it transpired, it was a sad series of events.

Compounding matters, Thai, one of three elephants Eloise had on tour with her, disappeared into the nearby forest to escape the chaos. "He was lying on the snow in the woods when I got there the next morning," Morgan told *The Oregonian* in 1978. "When I called to him, he staggered to his feet, stiff from the cold, and came running." Morgan later scattered Eloise's cremated ashes on Elephant Mountain.

I can't credit Morgan Berry enough for his pioneering work with elephants. His knowledge about the breed and their mating habits comforted me immensely during Belle's pregnancy. Packy's birth would never have gone as well without him.

For that, dear friend, I thank you.

Herd Of Wild Elephants Roams Through Tall Firs

Unique Hilltop Animal Farm Mixes Camels, Cows, Llamas

By LEVERETT RICHARDS
Staff Writer, The Oregonian

MORGAN BERRY HILL, Wash. (Special) — When you wake up in the morning on Morgan Berry Hill you're quite apt to find a llama in the living room, a wild cat in the kitchen and an elephant eating off the fireplace fender.

There will be a native deer peering in the window, to the mild amazement of the White Fallow deer on the other side of the fence. Gertrude, the hippo, will be making love to Tonga, the bull of the elephant herd. The eland will be ogling Shuroa, Mrs. (Jane) Berry's magnificent Arabian mare.

The Shetland stallion will be making love to Clyde, the (female) dromedary camel. The llama will be making love to the Jersey cow. And the leopard in the barn will be snarling his contempt of the whole love colony.

That's what it's like to be owned by nine baby elephants, one baby cheetah, one juvenile African antelope (eland), one lovesick hippo, two lonesome dromedary camels, five frisky horses, a compatible cow, three White Fallow deer, a llama that thinks it's people, and an assortment of bottle-fed baby animals that rises and falls like the tides.

BABY CHEETAH snuggles up on davenport in living room of hilltop house as Mrs. Morgan (Jane) Berry, herself a veteran animal trainer, tries to catch up on reading.

Morgan Berry's wife, Jane, often had her hands full with the many animals her husband brought home. Life, I'm sure, was never dull in the Berry household. (Reprinted with permission from The Oregonian*)*

A Zoo Ahead of the Herd

In 1959, one year after I came on board as its vet, the Oregon Zoo opened—to considerable fanfare—at its present location on Sylvan Hill's eastern flank, two short miles west of downtown Portland. It was, to say the least, a major improvement over the previous zoo, which was located where the Japanese Garden now resides in Portland's sprawling Washington Park. That zoo consisted of a dark, windowless, octagon-shaped building and a handful of outdoor enclosures that took maybe ten minutes to tour. I often told Jack Marks, the zoo director at the time, that the only thing holding that zoo together was the smell. The zoo troubled me so much, in fact, that I contemplated quitting. But I had made strides in improving the animals' health, especially with tuberculosis in primates, and I knew they'd be worse off without me.

The hard work of Jack Marks led to the construction of today's Oregon Zoo. It was a tough fight, however. Voters nixed increased taxes to fund the project until the zoo got its first elephant, Rosy, who instantly won local hearts. (Reprinted with permission from The Oregonian*)*

The new zoo reflected a changing attitude in America—and internationally—about how captive animals should be treated. Historically, zoos were designed primarily for human viewing, not for animal comfort. This ideology stretched back to 3500 BC, when emperors and the elite collected animals and kept them in menageries, mostly to flaunt personal wealth. But those collections, which also included deformed humans, were open only to society's upper crust. It wasn't until 1765 that the first public zoo opened in Vienna. Nearly 100 years later, in 1860, America's first zoo debuted in New York City's Central Park.

Portland's zoo, reportedly the oldest zoo west of the Mississippi River, got its start in 1888. The story goes that an English-born pharmacist named Dr. Richard B. Knight bought two bears, a grizzly bear named Grace for $75 and an Alaskan bear named Brownie for $50. The former sailor housed the bears and a host of other exotic animals in the back of his Southwest Morrison Street drugstore, and regularly displayed them on a vacant lot next door. Richard's wife, though, did

Richard Knight, a seaman turned pharmacist, bought animals from fellow seamen who traveled abroad and kept them in the back of his drugstore at Southwest Morrison Street near Fourth Avenue. (Courtesy of the Oregon Zoo)

not share her husband's enthusiasm for animals. According to the Oregon Zoo, Mrs. Knight considered the bears a "foolish extravagance" and, understandably, a danger to the couple's four young children. So in 1888 Richard asked the Portland City Council to buy the bears, using the argument, "They are gentle, easily cared for, and cost but a trifle to keep. And knowing they would prove a great source of attraction to the city park, [I] would like an offer for them before sending [them] elsewhere." Instead of buying the bears, the city gave him two circus cages and space to display them at the City Park (which became

R. B. Knight,

Druggist and Apothecary.

Full Line of Drugs, Toilet Articles, Shoulder Braces, Trusses and Elastic Hosiery, any size, Silk and Cotton.
Orders by Mail Attended to.

72 Morrison Street, near Fourth.

To — Portland, Oregon, 6/6/1888

The Honorable Mayor & City Council

Gentlemen

I have brought to this city and have for sale two bears, one young male brown, and a she grizzly, which latter is said to be with cub. They are gentle, easily cared for, and cost but a trifle to keep, and knowing they would prove a great source of attraction to the city park, would like an offer for them before sending elsewhere.

Yours respectfully

R. B. Knight

In 1888 Knight wrote a letter to the city, offering to sell his two bears, Brownie and Grace, to Portland. He eventually donated the bears, and the city's zoo was born. (Courtesy of the Oregon Zoo)

Bison grazed in their paddock at Portland's original zoo, which was located next to the city reservoirs in what is now called Washington Park. (Courtesy of the Oregon Zoo)

Washington Park in 1909). But five months later, Richard, who was still responsible for feeding the bears, finally just gave the animals to the city.

The original Portland Zoo was located above the city reservoirs that were built in 1884 to supply water to the growing city. Zookeeper Charles Meyers, a former seaman who had traveled extensively and observed zoos in foreign lands, designed and built the double-deck bear pit, believed to be the first sunken, barless cage anywhere in the world.

The zoo grew rapidly, and by 1894 the public could see more than 300 specimens, including a kangaroo. In 1905 Portland's Lewis and

Can you believe this? Rose Festival Princesses in the 1920s posed for pictures in the zoo's bear pit—unthinkable behavior at today's zoo. (Courtesy of the Oregon Zoo)

Clark Centennial Exposition left behind a lion, a polar bear, a leopard, several Olympic elk, and a pair of Yellowstone bison. The bison were kept in a gigantic paddock on the park's upper hillside, and "it was a common sight to see the bison grazing . . . above the reservoirs" (Lisa Patterson, *Oregon Journal*, 1971).

In 1925, however, the zoo was moved much higher up the hill in Washington Park, to the current Japanese Garden site. Built for $40,000, the new fourteen-acre zoo was, as Lisa Patterson puts it, "an animal Alcatraz." A few species, like the bears, had access to outdoor recreation, but the bulk of the animals were crammed into the large, octagon-shaped building. My wife, Patricia, visited that zoo as a child and distinctly remembers the horrific conditions. "The windows were so filthy from the dirt, feces, and age," she says. "My mother was

This picture, likely shot before 1925 at Portland's original zoo, shows an employee feeding the bears. (Courtesy of the Oregon Zoo)

horrified. We were only there for a short time because it seemed the animals were being mistreated."

Zoo attendance lagged, both because of the crummy conditions (you had to hold your breath to tour the main building) and the remote location. Unlike the old zoo, which was accessible by a walkway from Burnside Avenue, this zoo required a car ride up winding park roads that could be treacherous in bad weather.

The zoo did not improve much by 1958, when I came on board as vet; the city did not want to spend the money, especially during the Depression and World War II. The animal population then numbered 250 and featured lions, black bears, polar bears, monkeys,

2* Oregon Journal SATURDAY, DEC. 24, 1955

STORMS MAY COME and storms may go, but they'll not stop stork from keeping appointments. As proof of this, llama population of Portland zoo in Washington park was boosted by one, shown by photo of mama llama, youngest.

Before the current zoo opened in 1959, conditions for animals at the old zoo were less than ideal. Still, animals continued to be born and celebrated, like this baby llama that arrived in 1955. (Reprinted with permission from The Oregonian*)*

birds, bison, two camels, and two elephants. Keepers were city workers who had failed at their other jobs; working at the zoo apparently required zero skills, according to the city. Actually, keepers did need one certain skill: the ability to move fast. Zoo cages were not designed to keep animals away while keepers cleaned. Worse, many of the steel enclosure gates had rusted shut. Keepers literally had to climb over fences to clean, keeping one eye on the bears or lions that might charge at any moment, especially during mating season.

Jack Marks deserves much credit for the construction of the new zoo. Hired as director in 1947, Jack tirelessly pushed for a better zoo, even contributing to its animal population. Three times, in 1957, 1958, and 1962, Jack flew to the Antarctic to capture penguins for Portland's zoo and elsewhere.

THE OREGONIAN, WEDNESDAY, MAY 16, 1956

This Baby Bison Is New at Zoo

Bison grazing on the slopes of Washington Park were a common sight for Portlanders in the 1950s. Frisky, this baby bison, joined the zoo's herd of seven buffalos in May 1956. (Reprinted with permission from The Oregonian*)*

In 1958 thirty-eight of Jack's penguins lived in a North Portland pool until the new zoo's aquarium was built. We drained and scrubbed the pool daily and took other precautions to ensure a bacteria-free environment, but several penguins still developed a fungal lung disease called aspergillosis. That disease, common in all bird species, killed several of our penguins and got me working on a cure. It took a couple of months to figure it out, but I

Jack Marks should have been an explorer. For more than twenty years he steered Portland's zoo, but I bet his three penguin-finding trips to Antarctica thrilled him most. (Courtesy of Marianne Marks)

finally pioneered a misting technique that conquered the problem. To administer the treatment, the penguins were corralled into a contained room and cloaked with a fog of Amphotericin, an antifungal antibiotic. This method proved so successful that the poultry industry used it to prevent aspergillosis in its birds. Hospitals also employed this method to combat disease and to sterilize contaminated equipment.

Even before Jack's penguin-finding journeys, he campaigned hard for the new zoo. It proved a tough fight. In 1951, backed by the City Club of Portland, Jack convinced city leaders to put a $3.85 million bond issue on the ballot for a new zoo. But debt-wary voters rejected it.

Jack didn't give up. He kept lobbying, and soon he had a secret weapon. In 1953 Oregon attorney Austin Flagel, on an economic mission to Thailand, bought a Cambodian elephant named Rosy to give to Portland's zoo. Oregon's first-ever elephant, considered royal because of her pure white skin, so enchanted Portland schoolchildren that they raised the money to pay for Rosy's shipping costs. Rosy equally charmed voters, who in 1954 passed the previously defeated zoo bond measure.

Portland got a second elephant in 1956, about the time construction got underway on the new zoo. Portland native Orville Hosmer, an engineer with a United States Operations Mission in Vietnam, received a seventeen-month-old female elephant named Tuy Hoa (pronounced "TEE-waa," meaning "peace") from the grateful Vietnamese people of the Tuy Hoa Valley. Orville had supervised the restoration of a local irrigation system that had been destroyed in the early 1950s during the French Indochina War.

City leaders decided that the new zoo would be built on the western backside of Washington Park, on acreage that once housed the Multnomah County Poor Farm and, more recently, the nine-hole West Hills Golf Course. Although accessible from several directions, the main entrance to the zoo was off of Canyon Road, which today is the Sunset Highway.

Construction began with boasts that this zoo would be "one of the most interesting, pleasing, and outstanding" zoos in the world, reported *The Oregonian*. Architects Abbott Lawrence and Ernest F. Tucker, accompanied by Portland's park superintendent Harry Buckley, toured zoos nationwide to glean the most modern thinking in zoo layout and construction. Ultimately they decided to capitalize on Sylvan Hills' sloping vistas, which allowed animals to be grouped as they might appear in nature—as opposed to creating elaborate scenes of their natural habitats.

"We found that the animals are better seen in simpler backgrounds and in moated grottoes of moderate size," said Abbott Lawrence,

In 1955, bulldozers shaved the fifty-acre zoo site after loggers chopped down about 200,000 board feet of timber worth $16,000 (in 1955 dollars). The huge grading contract required the moving of some 200,000 cubic yards of dirt. (Reprinted with permission from The Oregonian*)*

"which bring the animals as close to the spectator as is safe" (*The Oregonian*, 1957).

That's not to say that the elephant exhibit offered the same space as the jungles or open plains of Asia and Africa. But the spacious yards and soaking pool provided the best home-away-from-home that an elephant could have at that point in zoo history. And the elephant barn showcased safety features like hydraulic doors that enabled handlers to safely separate the big animals—especially cranky Thonglaw, when he was in musth.

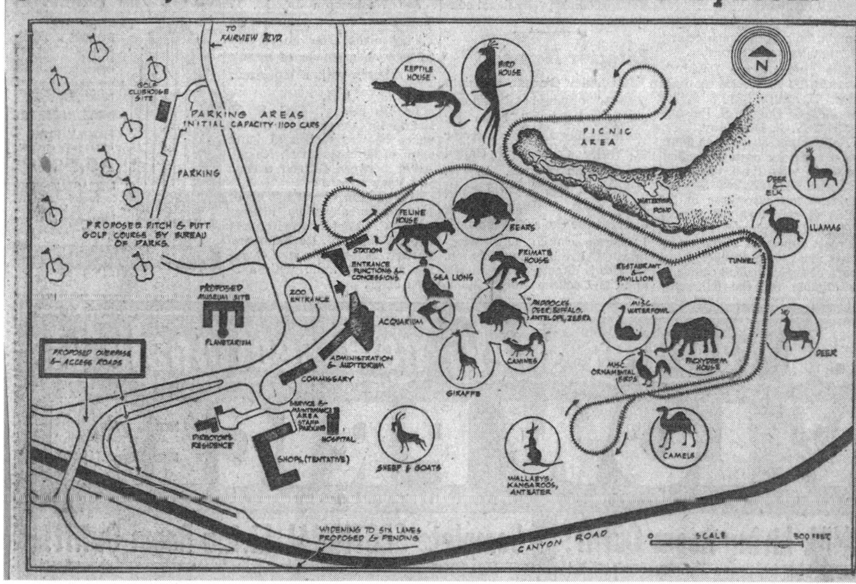

This illustration, published in 1955, shows the proposed route of the zoo's original "kiddie" train. The one-mile journey, which provided riders with multiple views of the zoo, looped past camels, kangaroos, and elephants, just to name a few. To help pay for the zoo line, schoolchildren sold "stock" at $1 per share, and a children's book entitled Clickety Clack and the Bandits *was created and sold. The Portland Zoo Railway began operation on June 9, 1958, with the Zooliner, which is still the primary train of today's Washington Park and Zoo Railway. The expanded route now features three trains, including a Washington Park Run that chugs four miles roundtrip through the park. The railway carries more than 350,000 passengers annually. Notice the proposed pitch & putt golf course at left in the location of the current Portland Children's Museum and World Forestry Center. (Reprinted with permission from* The Oregonian*)*

It Was Elephant Moving Day at Portland's Zoo

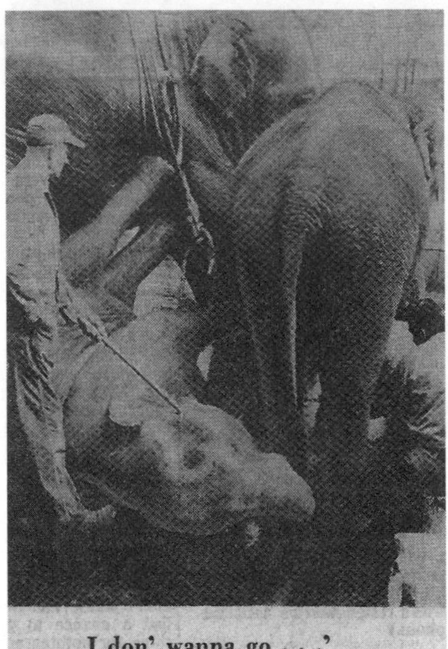
I don' wanna go . . .'

Then we'll drag you . . .

Get up, you little *x! . . .

Now you're being sensible.

I much admired the new elephant enclosure, but I also wished planners had consulted me about its design. In the wild, elephants plod hundreds of miles in search of food and water, a process that naturally maintains their huge feet. In zoos, where walking is limited, it's important that the elephant enclosure and yard be surfaced with material that cushions their feet. Because this didn't happen until decades later, several elephants—Belle, Thonglaw, and Pet among them—suffered severe foot ailments.

It was a giant undertaking moving animals between the old and new zoos, but we did it in about two weeks. Some critters traveled in crates, but many animals like the elephants were herded on the winding roads through Washington Park. As newspaper photos from the time attest, elephants don't always cooperate! Sometimes they lay down and sometimes they stood like statues, unwilling to budge. Keepers had to tug, push, and goad to finally get the massive animals to their new quarters.

Tragically, we could not move the bears because of the danger they presented. Sedating drugs, still in development, were not used extensively, and the bears refused to be caged. In the end, we had no choice but to shoot and kill them—cruel by today's standards but necessary. In childhood I killed animals for the dinner table but never for sport. The one time I hunted for pleasure I was so troubled by the experience

(Opposite) In 1959, animals were moved from the old zoo, where the Japanese Garden now sits, to the present zoo's location. We herded the big animals on a windy road through Washington Park, and Morgan Berry brought in Thonglaw to help guide Rosy and Tuy Hoa to their new home. Progress was slow, however, as shown in this series of four photos. (Reprinted with permission from The Oregonian*)*

This aerial photograph, shot in 1959, shows the layout of the new zoo when it opened. Much of the zoo, including the Children's Zoo, was not finished at this point. The oval-looking elephant exhibit is located in the bottom/center. (Reprinted with permission from The Oregonian*)*

that I never did it again; I'm equally disturbed each time I must put an animal down. Patricia says I retreat within myself, to a safe place where I am able to do what needs to be done. Then it always takes my heart and my head awhile to recover. But when zookeepers repeatedly shot at the bears, wounding them horribly but not putting them down, I took the rifle into my own hands and swiftly finished the gruesome chore.

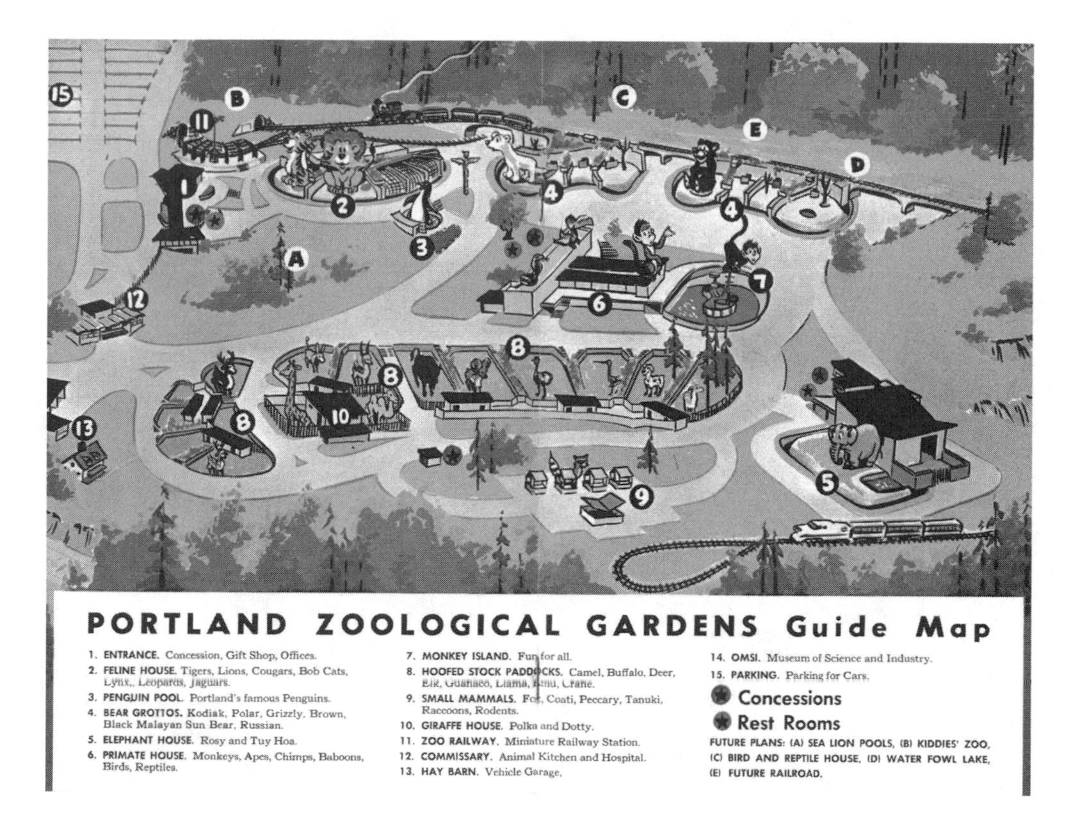

This map was given to visitors of the new zoo, so they could find their way to various exhibits. (Courtesy of the Oregon Zoo)

Aside from these tragedies, the zoo's grand opening in July 1959 (which coincided with Oregon's centennial) was a major success. All of the animals had a much better living situation, pleasing me tremendously. I tip my hat to Jack Marks and all the others who helped to create this new zoo—and Packy's lifelong home.

Folks, This Is No Peanut Announcement

Portland Zoo's Three Elephants Are All Expecting

Associated Press.

PORTLAND, Oreg., Jan. 10. —The secret is out: Three elephants at the Portland Zoo are pregnant.

This is a momentous event in zoological circles. Available records indicate that only six elephants have been born in this country previously, all between 1880 and 1919.

The expectant mothers will contribute to medical science. It seems that no one knows the length of the elephant's gestation period. It has been variously estimated at 21 to 24 months.

Morgan Berry of Seattle, owner of one of the pregnant elephants, explains: "No accurate records have been kept in the jungles and there has been no chance to keep records in zoos. Dr. Matthew Maberry, the zoo veterinarian, will be the first to keep exact, certified records and publish them."

Dr. Maberry says the first of the elephants to deliver will be 10-year-old Belle. He predicts she will give birth sometime after next week. If Dr. Maberry is right, this will establish the gestation period at 18 months.

Also expecting are Rosy, age 12, and Tuy Hao (pronounced Tee Wa), who is 8. Rosy's baby is expected in July and Tuy Hao's around September. If Tuy Hao delivers, says zoo director Jack Marks, she'll be the youngest elephant to give birth in captivity.

Father of the impending brood is 14-year-old Thonglaw.

Dr. Maberry gave this report on his patients.

Belle has gained more than 800 pounds in the past year and weighs 2½ tons. She seems restless and has some odd tastes.

Rosy, the oldest, weighs three tons and she is miserable. She's been that way from the first month.

"She has morning pains that last all day," Mr. Marks explain.

Tuy Hao, whose weight is a trim two tons, feels fine.

Officials at the Bronx Zoo in New York said their records list six elephants born in this country between 1880 and 1919.

William Bridges of the Bronx Zoo said such births are rare, not because of difficulty of mating elephants in captivity, but because few zoos or circuses keep males — they tend to be savage.

"Portland," he said, "has a very fine male, however, and ought to be pleased if the births occur as expected."

I shocked the public in January 1962, announcing the birth of not just one baby elephant but three! (Reprinted with permission from The Oregonian*)*

January 1962:
The Countdown Begins

I confirmed Belle's pregnancy in December 1961, but lacking a due date, Morgan Berry, Jack Marks, and I conspired to keep the pregnancy a secret until the blessed event. On January 18, however, when Belle started having intense labor pains and birth seemed imminent, we decided to break the big news.

Well, that did it. All heck broke loose, especially when we revealed the *really* big news: not only was Belle pregnant and likely to deliver any day, but Rosy and Tuy Hoa were also expecting. (I suspected young Pet was among the mothers-to-be, but I saved that announcement for later.)

I've never seen such a public frenzy in my nine long decades of life! Newspapers, televisions, and radios all trumpeted the big news as if it were First Lady Jackie Kennedy giving birth. Suddenly low-profile Portland had something to both shout and gloat about, and

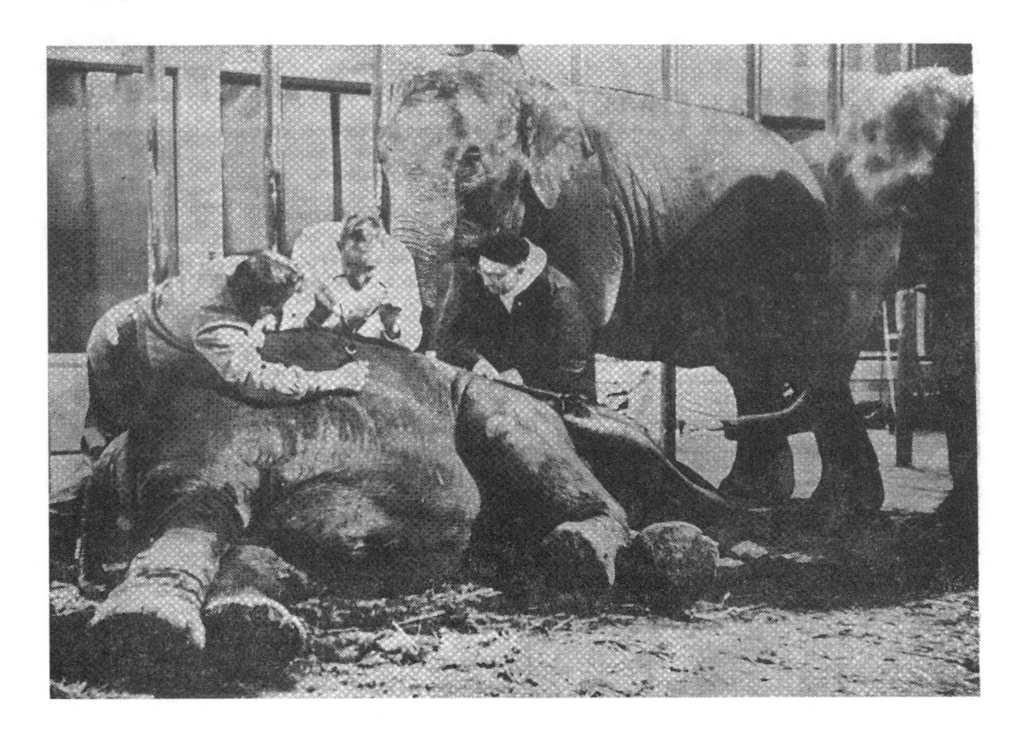

In this picture Dr. James Metcalfe, then an associate professor of cardiology at the University of Oregon Medical School, center, and Morgan Berry, right, helped me once again check the pregnant pachyderm's condition. Of course, the other elephants wanted to know what the heck was going on. (Reprinted with permission from The Oregonian*)*

the city—no, make that the entire state—literally quivered with anticipation. Nothing like this had happened in Portland's history (or world history, for that matter) and it felt good, wholesome. Blessedly welcome, especially to adults, distressed by the icy Cold War and crazy social changes bucking America's *Father Knows Best* backbone.

In 1962, while some citizens dug bomb shelters in their backyards and stocked them with enough food to ride out an A-bomb, I was thinking babies. Belle was due anytime, Rosy would drop sometime in fall 1962,

The two-story farmhouse where I spent the bulk of my childhood and early adulthood still stands in Sequim, Washington. (Courtesy of Dr. Matthew and Patricia Maberry)

and Tuy Hoa, I guessed, would deliver one year later. Add Pet's likely pregnancy to the mix, and I had myself one full maternity ward.

I also had one grouchy father on my hands. In Belle's final months, Thonglaw entered musth and had to be isolated. He rattled everyone, including *Oregonian* reporter Lev Richards, who wrote, "From time to time, he lunged at the two-inch bars on his isolation cell, trying to crush [Morgan] Berry, who warily eluded his rushes. Each charge shook the building, waking some of the attendants and [news] press who slept fitfully on a pile of baled hay in the feed room during the long vigil of the night."

Like I said before, knowing Belle's due date would have been a huge help. It really bothered me that I didn't have many answers at all, as I was raised to find them. My childhood farmhouse lacked urban luxuries like running water, but my parents, Walter and Bertha, were avid learners who listened to classical music and maintained a well-stocked library. We descend from

My father, Walter Eugene Maberry, believed that hard work molded a man's character and books shaped his mind. His untimely death in 1939 caused me deep sorrow—so much that I left veterinary school for several years to work at home on the farm. (Courtesy of Dr. Matthew and Patricia Maberry)

a long line of big thinkers—doctors, lawyers, botanists—among them, John Bartram, the father of American botany. His forty-five-acre homestead on the Schuylkill River in southwest Philadelphia, now a national historic site, features his famed botanic garden; my middle name, Bartram, is a nod to this legacy. Growing up, in fact, most folks called me Bartram. And whenever I asked my father a tough question, he maintained our lineage with his stock response: "Let's look it up."

My father also taught me to expect the unexpected, since life often surprises. His life, I think, was one surprise after another. Born in Arkansas in 1865, the year Abraham Lincoln was assassinated, my father's family traveled west by covered wagon in 1871. My father, though, got sick somewhere around modern day Montana, and his family had to let the rest of the wagon train go ahead. Native Americans helped my father heal. Finally, after years on the Oregon

My mother, Bertha Belle Veihmann Maberry, was a remarkable woman. Our farmhouse lacked electricity and running water, yet she always made sure my father and her four boys had a clean, pressed shirt to wear every day. (Courtesy of Dr. Matthew and Patricia Maberry)

Trail, the family reached Portland in 1882, but my grandfather disliked the city, so they backtracked to Idaho. It was here that my father, at age nineteen, bettered his third grade education by hiring a tutor to teach him engineering. He got a job supervising an Idaho quartz mine, but in 1900, at age thirty-five, my father (like thousands of other ambitious American men) abandoned his career to pan for gold. From Jacksonville, Oregon, to nippy Nome, Alaska, his gold quest ultimately took him to the Tanana River in central Alaska, where he stayed for a decade.

Eventually, my father figured he'd better get out of Alaska if he ever wanted a family. Around 1910 he returned to the lower 48 and settled in Seattle; that's where he met my Philadelphia-born mother, Bertha, who was decades younger than him. They married and had five children—four boys and a girl; I'm the middle child. Then my father

Here I am with my three brothers and my sister. In the back row, from the left, are Walter, Jim, and me, while Betty and Merrill are in the front row. Unfortunately, I'm the lone survivor of all my siblings, and I sure do miss them. (Courtesy of Dr. Matthew and Patricia Maberry)

convinced my city-raised mother that if they were going to raise us boys right, they'd better live on a farm. Looking back, I can see his point. But growing up, saddled with never-ending chores, I thought farm life was going to kill me.

My father taught me a lot about animals, not just the ones in our barns but also the squirrels, deer, and quail that darted about the wildflower meadows surrounding our farm. We took many long walks there, in Sequim's Happy Valley, talking about everything we saw. Even then, I realized his life's depth: the Oregon Trail, Native Americans, the Alaska Gold Rush. He made me want to be him.

Every year I usually raised a 4-H animal for competition at the county fair. Here I am, probably around age ten, with a cow I took care of from birth. (Courtesy of Dr. Matthew and Patricia Maberry)

I wish my father had lived to 1962 so he could have watched the birth of a baby elephant. Sadly, he died in 1939 when a horse kicked him in the head. But he left me well-trained and brimming with common sense. We never had a vet visit our farm; none practiced nearby. My father, in collaboration with the effort and know-how of our neighbors, doctored our 100 cows, a handful of horses, and an assortment of other critters. Nothing on the farm came easy, but I think the lifestyle instilled in me a keen appreciation of life and a deep empathy for animals.

In January 1962 I felt that same deep empathy for Belle, whose enormous body was regularly wracked by false labor pains; she convinced me several times that she was hours away from delivery. Consequently, I learned that it takes weeks, even months, for a baby elephant to move into position for birth. She would weave steadily, fore and aft, as if she were rocking her 200-pound baby to sleep. Sometimes Belle became bug-eyed and her forehead swelled; one time she whirled and, with startling speed, seemed to charge me. Other times she tried to stand on her head.

I can only imagine how Belle felt. She had never seen a pregnant elephant before, and Packy regularly pummeled her sides, which zoo staff and the media—all members of the Great Portland Elephant Watch—could clearly see. We truly felt for her; weight gain, a kicking baby, and intense labor pains no doubt perplexed the mother-to-be. To give you an even better picture, an elephant in labor exerts about four times the force of a horse in labor, and Belle's pains seemed gigantic. Sometimes, when she shrieked in pain and tears rolled down her trunk, Pet, Rosy, and Tuy Hoa crowded close and cried in empathetic agony. Several times Pet used her knee to massage Belle's belly; twice she tried sitting on Belle's head, perhaps to silence her agonized cries.

Whenever Belle began to deliver, I planned to be ready. Beyond my surgical instruments, I had a resuscitator, a respirator, and special 110-volt electric prods that I could use to stop Belle if she tried to trample her baby. I also had floodlights to illuminate the birthing area, and I borrowed a set of shipping room scales to weigh the newborn. I tried to anticipate every birth scenario, including Caesarean section. I had performed C-sections on cattle and horses, but I truly hoped a C-section wouldn't be necessary, as I had *no* idea how much anesthetic it would take to sedate an elephant. More troubling, I wasn't sure how I would strap a makeshift gas mask on Belle's trunk. I had an oxygen tent, but it wasn't big enough. A pup tent would work, but it wasn't airtight.

I also sought advice from two physicians who specialized in humans: Dr. Howard Tatum, professor of obstetrics at the University of Oregon

Weary Zoo 'Elephantricians' Wait, Keep Alert For Overburdened Stork

By LEVERETT RICHARDS
Staff Writer, The Oregonian

Pictures On Page 8

The stork remained grounded by bitter weather Friday and was unable to keep its date with Belle, expectant elephant at the Portland Zoo.

The two-and-a-half tons of pregnant pachyderm seemed restless and undecided all day Friday as the zoo attendants kept vigil.

Dr. Mathew Maberry, zoo veterinarian late Friday called a conference with Dr. James Metcalfe, associate professor of cardiology, and Dr. Howard Tatum, obstetrician from the University of Oregon Medical School, on the perplexed pachyderm's puzzling pattern of behavior. The motherly elephant owned by Morgan Berry of Seattle had her first gigantic labor pains Thursday afternoon and has been uneasy ever since.

The "elephantricians" decided to consult the only living man known to have delivered an elephant outside of the jungle countries, Dr. Eramamno Brozini, of the Rome Zoo. Dr. Brozini presided at the birth of a healthy baby in the Rome Zoo Sept. 28, 1948.

Call Attempted

An attempt was made to put through a long distance call Friday night with the aid of an Italian interpreter. Veterinarians and other doctors are particularly interested in Belle's impending blessed event because of the scarcity of scientific information on elephant birth.

Dr. Brozini could not be reached, but Dr. Maberry put through a conference call to Dr. Emmanuel Amoroso, of the Royal Veterinary College of London, who had seen many elephants born in jungle work camps. The consultation, courtesy of the Pacific Northwest Bell Telephone Co., recorded through KOIN's switchboard, confirmed most of Dr. Maberry's diagnoses.

He believes Belle is just starting into labor, which lasts from 6 to 12 hours, sometimes longer for the first baby. The baby is born head first while the mother squats. She may be helped by her attending females. She ties the umbilical knot herself.

The heart beat of the infant, as heard by Dr. Metcalfe, a cardiologist, shows the baby in position for birth and discounts the possibility of twins. The baby comes swiftly once started, Dr. Amoroso warned.

With that word from the wise Dr. Maberry, still the only living American to play midwife to an elephant, turned in for the night in the elephant barn, confident a new voice would be heard in the elephant house before the week ended.

As this story shows, I resorted to every source possible to find the elusive answer to Belle's delivery date, including officials at the Rome Zoo, where a captive elephant had recently been delivered. (Reprinted with permission from The Oregonian*)*

Reporters turned to clichés, like the stork, to provide buildup in the days, weeks, and months leading to Packy's birth. (Reprinted with permission from The Oregonian*)*

Wing-Weary Stork Sets Approach Pattern At Zoo

By LEVERETT RICHARDS
Staff Writer, The Oregonian

Picture On Page 31

The stork, with sagging wings, hovered low over the elephant house at Portland's zoo Sunday morning, waiting to deliver the first baby elephant born in the U.S. in 43 years.

Belle, the 10-year-old Indian elephant which has been laboring mightily to deliver the long-awaited infant, was in the final stages of labor at 1 a.m., Dr. Mathew Maberry, zoo veterinarian, reported.

Two consulting physicians agreed. Dr. James Metcalfe, cardiologist, and Dr. Howard Tatum, obstetrician, from the University of Oregon Medical School, remained at Belle's bulging side throughout the night.

Baby Seems Restless

The pulsating pachyderm was having regular contractions, visible on her heaving sides, by 10 p.m. She was "weaving" steadily, fore and aft, as if to rock her restless 200-pound baby to sleep. Heartbeats read by Dr. Metcalfe and examinations by Dr. Maberry indicated the baby's head was well up in her mother's pelvis, ready for normal, and probably precipitous, delivery.

All was quiet at midnight as Tuy Hoa and Pet, Belle's attendant "midwives" lay down on their sides to sleep the sleep of the weary. P.S. Elephants do snore after all.

But Rosy, also pregnant, stood solicitously by Belle throughout the night.

Belle Shows Strain

While Belle, an unusually affectionate elephant, continued to demand attention from her attendants, she showed signs of a possibly dangerous change in temperament. She became bug - eyed, her forehead swelled, and on one occasion she whirled with startling speed and seemed to charge toward her attending physician as he stepped up behind her.

Dr. Maberry was inclined to believe, from all the symptoms, that the baby would be born before dawn, Sunday, 18 months and 2 days after conception. It may be sometime before the sex of the new arrival can be determined, he advised. It will be at least two days before he can take the baby away from the mother even long enough to weigh it, he said.

Rosy and Tuy Hoa entertained visiting elephants Belle, Pet, and Thonglaw in the elephant exhibit's swimming pool during the never-ending pregnancy. (Reprinted with permission from The Oregonian*)*

Medical School, and Dr. James Metcalfe, associate cardiologist at the same medical school and president of the Oregon Heart Association. Neither knew anything about elephant birth, but they became the first ever to record the heartbeat of an elephant in utero. After handlers coaxed Belle to lie on her side, the doctors wrapped a canvas fire hose—entwined with wire—around her girth. The wire connected to an electrocardiograph machine that recorded the heartbeats of both Belle and Packy. Belle's fifty-pound heart maintained a steady forty beats per minute, while Packy's heart thumped sixty-eight beats a minute.

The study inspired me to help Portland businessman Loren Parks invent a transmitter that allowed the electrocardiograph findings to be

A medical team from the University of Oregon wanted to record the heartbeats of Belle and her unborn baby to see what they could learn. Here Dr. James Metcalfe, left, fed Belle peanuts while I adjusted the straps on Belle's belly and Morgan Berry, on the floor, adjusted the electrodes needed for the electrocardiograph machine to record heartbeats. (Reprinted with permission from The Oregonian*)*

recorded outside of the elephant barn. I worried that Belle might kick or otherwise disrupt the delicate equipment used to make the recording. In coming years I used this technology repeatedly, with elephants and other animals, to help detect a second heartbeat and thereby confirm a pregnancy. I was happy with the new technology, but it did little to enlighten me on a possible due date.

No, this time the answers weren't something I could look up in a book. Somehow, however, I knew everything would be okay. I've never been an outwardly religious man, but I like to think I have the good Lord on my shoulder.

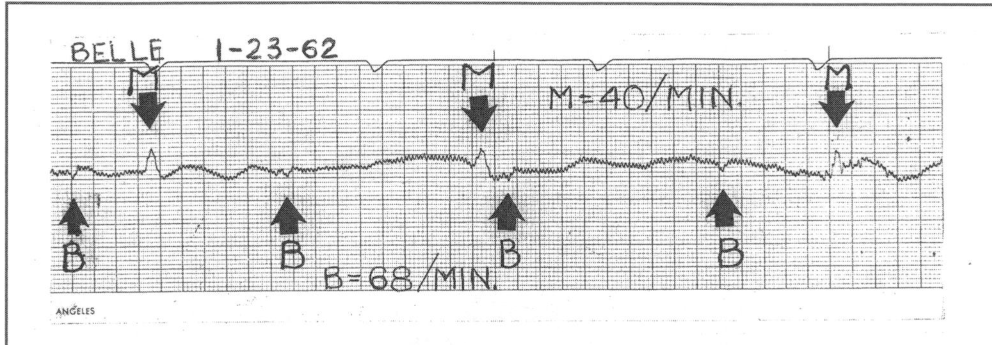

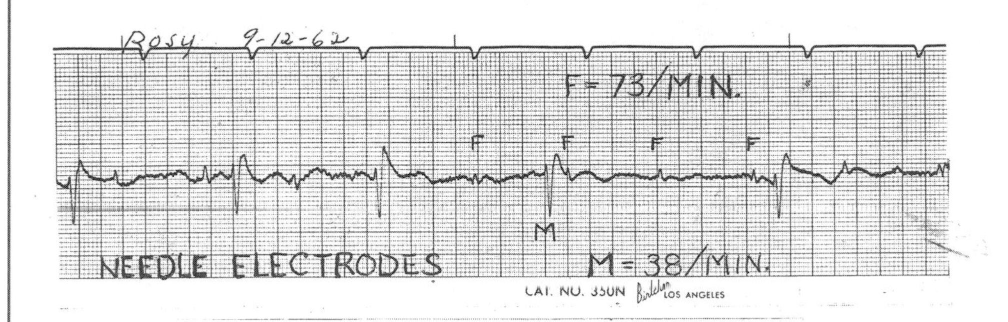

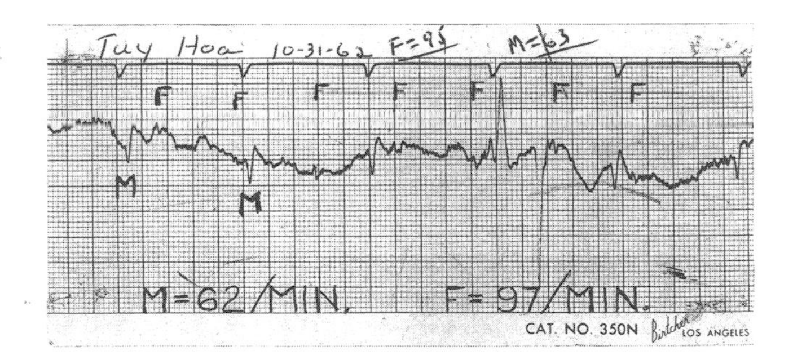

These electrocardiograph readings, recorded while the three mama elephants were pregnant, show the heartbeats of Belle and Packy, top; Rosy and her baby Me-Tu, center; and Tuy Hoa and her baby Hanako, bottom. Belle's reading, for instance, showed her heart pumping at forty beats per minute, while Packy's thumped at sixty-eight beats a minute. (Courtesy of Dr. Matthew and Patricia Maberry)

Many Offer 'Advice' To Zoo Vet

By JACK VANDER WHITE
Journal Special Writer

Belle's Calf Key To Pool

Wing-Weary Stork Sets Approach Pattern At Zoo

By LEVERETT RICHARDS
Staff Writer, The Oregonian
Picture On Page 31

The stork, with sagging wings, hovered low over the elephant house at Portland's Zoo Sunday morning, waiting ...

... on their sides to sleep the sleep of the weary. P.S. Elephants do snore after all.

But Rosy, also pregnant, stood solicitously by Belle throughout the night.

Belle Shows Strain

While Belle, an unusually affectionate elephant, continued to demand attention from her attendants, she showed signs of a possibly dangerous change in temperament. She became bug-eyed, her forehead swelled, and on one occasion she whirled with startling speed and seemed to charge ...

Belle Keeps Vet Waiting

Belle, Portland's expectant elephant, was still expecting Monday, 11 days after her first signs of labor, with no delivery date yet in sight, Dr. Mathew Maberry, zoo veterinarian reported.

More than 5,000 ... ers flocked to the z... most of them after ... tend their sympath... on-again-off-again ... This would be a g... even in summer, Dir... Marks pointed out.

Belle Continues To Mark Time

Stork Stalls Belle Birth

A watched pot never boils, as everybody knows. And Belle, the pot-bellied ... the Portland Zoo, wa... to boil Tuesday.

Her huge sides ... tight as a drum. ... like she had swa... blimp. But she s... signs of resuming ... that apparently sta... bellow last Thursda... gressed to an anti-... urday night.

She rocked and r... mittently, but lay ... sound night's slee... morning just after ...

Baby Elephant? Nope, Not Yet

Belle, the absent-minded elephant, Wednesday still showed no signs of remembering she is supposed to be having a baby.

Zoo Belle Still Lags

Belle, the Zoo's reluctant mother, was still keeping her fans waiting for the birth of her baby Thursday.

Dr. Mathew Maberry, zoo veterinarian, reported his patient increasingly restless and ... She won't eat ... and dawdles over her ... tonic of sulphur and mo...

Calm Prevails On Zoo Front

A truckload of Timothy hay was delivered to elephant Belle at the Portland Zoo Friday to encourage the pregnant animal to deliver her calf.

A gift from the merchants of Eastport Plaza, the protein feed will reportedly ... at a ceremony Tuesday.

Belle Balks At Big Event

Belle, the absent-minded elephant, Wednesday still showed no signs of remembering she is supposed to be having a baby.

She seemed to be enjoyin... all the attention, but show... no major signs of resum...

Belle Still Waits Reluctant Stork

Yes.
No.
Yes.
No.
And "No" was the latest word Monday morning at the Portland Zoo as Dr. Jack Maberry and assorted assistants hopefully awaited the arrival of the first elephant calf to be born in captivity in ...

Belle Stalls On 'Duty'

Belle Frid... no hurry ... Dr. Mathew ... rinarian, re... The baby, ...

Zoo's Operation Stork Winging On Slow Belle

By LEVERETT G. RICHARDS
Staff Writer, The Oregonian

All the world loves a mother — especially if she's an elephant.

Belle, who will be ... mother of the ...

The Oregonian switchboard was busy answering elephant calls all day and ... night, an esti...

'Mommy' Elephant 'Ready'

An elephant named Thonglaw paced the floor at the Portland Zoo Friday.

Belle, a 10-year-old elephantine cutie from Siam, was expected to make him a father before the day was over. She was in labor throughout the night Thursday, indicating she soon would be the first elephant to deliver a baby in the U.S. in 43 years.

Standing by to assist in the delivery was Dr. Matthew Maberry, zoo veterinarian; Morgan Berry, Belle's owner; Zoo Director Jack Marks and his aides.

ANIMAL obstetrics is certainly not unusual at the zoo, but to deliver an elephant is another story. However, Dr. Maberry is confident there will be no trouble with the birth.

"Nature will take its course," was his philosophical comment.

The veterinarian said that as the labor pains progressed ...

Expectant Elephant Still Expecting

Belle Basks In Sun

Belle, Portland's expectant elephant, was still ...

Even Thonglaw, the bull, father of Belle's, Rosy's and Tuy ...

... PROUD OF BABY ELEPHANT

PORTLAND, Ore. (UPI) — Portland's famous newly-born elephant—tipping the scales ... 225 pounds— ... and tried to ... to do with ...

... grinned, ... mother, Belle, ... of the first ... born in this ... years.

... son peered from between ... but Belle ... to maneu... rarer view.

... flocked to ... Elephant ... the youngster ... med. They lined the ... wall of the exhibition room several deep, and were reluctant to move on.

Belle 'Rings' New Alarm

For five days now three men and four elephants have paced the maternity section of Portland Zoo's elephant house.

But nothing has happened. Belle, the 10-year-old expectant elephant, appeared close to delivery Tuesday morning as she has many times since the 24-hour baby watch began last Thursday.

But it may be another false alarm.

On hand are zoo director Jack Marks, and veterinarian Dr. Mathew Maberry, Norman Berry, of Seattle, owner of the elephant, has spent most of his time at the zoo.

Belle Fails To 'Ring'

Belle, the procrastinating pachyderm, was still procrastinating Tuesday.

She had further labor pains Monday night, Dr. Mathew Maberry, zoo obstetrician, reported, but failed to follow through. Her appetite is finicky and freakish. She toys with her hay, but greedily guzzles molasses water, which Dr. Maberry has prescribed for a tonic.

She and Rosy, Tuy Hoa and Pet are taking the air every afternoon during the current "heat wave," Director Jack Marks reported.

Articles abounded in Oregon newspapers in the weeks leading up to delivery, informing readers about anything and everything but mostly saying: Waiting, waiting, waiting. (Reprinted with permission from The Oregonian*)*

February 1962: Questions, Questions, Questions

Growing up in a small town, I certainly wasn't a celebrity. Paparazzi never hung out in our barns or trailed me to the nearby creek, where I cooled large trays of freshly squeezed milk (we skimmed cream off the top and sold it for fifteen cents a pound). I never worried how I looked in case someone snapped a photograph—heck, I rarely had my picture taken in my youth! Folks (my family included) were too busy eking out a living during the Depression.

The Great Portland Elephant Watch, though, instantly turned me into a celebrity of sorts. Suddenly everybody wanted my opinion, my thoughts, my wisdom. I became the guy with the crystal ball who would tell the world when Belle's baby would appear. Never mind that I didn't know. As the zoo's veterinarian, I was the person closest

Dave Falconer, a photographer for The Oregonian, *dodged a curious trunk poking from the elephant enclosure. Behind Dave, KGW newsman Dick Ross is interviewing me. (Reprinted with permission from* The Oregonian*)*

to Belle and the goings-on in her belly. Like it or not, that made me—and my good pal, Morgan Berry—the experts.

Writers, photographers, and reporters representing newspapers, magazines, television, and radio all planted roots in the elephant barn that winter and did not uproot until Belle gave birth. Every time I went anywhere near the expectant mother, at least a dozen pairs of eyes

Reporters, Photographers Keeping Nervous Vigil Over Expectant Belle

Lev Richards, a reporter for The Oregonian, *and the rest of the press pool practically lived at the elephant barn for more than two months. To pass the time, the media formed their own secret society called SPOPE: Society for the Preservation and Observation of Pregnant Elephants. (Reprinted with permission from* The Oregonian*)*

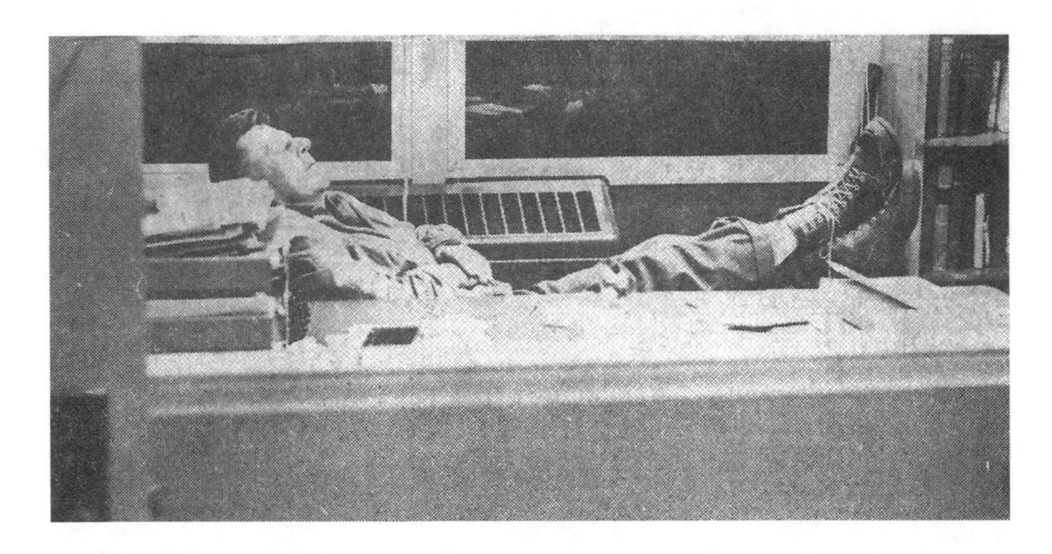

I had to sleep whenever I could catch a couple of winks during the long weeks leading up to Belle's delivery, and it seemed a camera was always there. (Reprinted with permission from The Oregonian*)*

trailed me. Pens and pads whipped out, television cameras whirred, radio microphones emerged. Questions crackled the air.

Even a routine chore like measuring Belle's belly somehow became big news—maybe because I used a fifty-foot measuring tape. Listening with a stethoscope for Packy's heartbeat also drew attention. It proved challenging, too, given Belle's one-inch-thick skin. But I listened hard, and pretty soon I could pick up two heartbeats.

All the attention got a little old. Up to then, I more or less worked in obscurity, making my daily rounds and caring for animals with little fanfare. With Belle, I always shared new information when I had it, but her condition—except for the occasional bout of false labor—changed little day to day. Imagine a gaggle of press types hanging out in a pregnant woman's living room, demanding to know details of her every twitch. Belle couldn't speak for herself, so I became her voice. But after the first few hundred questions, I wished everybody would go home and just let nature take its course.

That didn't happen. The public, ecstatic about this elephant birth, demanded every morsel of pregnancy trivia. Radio stations broadcast hourly "Belle-tins," but that wasn't enough. Curiosity abounded: "How does an elephant cut and tie an umbilical cord?" "How does an elephant dry its newborn babe and wipe off the protective mucous coating?" "Is a baby elephant born trunk or tail first?"

Reporters tried to answer every question, but not every question had an answer; too little was known about captive elephant birth. So, in fun, the media turned to nontraditional sources, like psychics. Clairvoyants, though, proved no smarter than me on the delivery date. As one psychic put it, "One would have to know the circumstances of the mother's birth—the time figured on Greenwich scale when she was born and the atmospheric pressure of the day, and the influences of stars and planets." *Oh brother* is pretty much what I thought about that revelation.

Oregon stores sold tons of elephant-related merchandise, especially stuffed animals. A florist created a gigantic papier-mâché bootie and filled it with 300 long-stem roses; the elephants skipped the fragrant petals and munched instead on the thorny stems. Gifts poured into the zoo—everything from gold-plated diaper pins to an elephantine-sized diaper fringed with frilly pink lace and rosebuds, courtesy of the Grandmothers Clubs of Oregon.

Schoolteachers capitalized on the birth as well, shaping their lessons around elephants. One teacher, in fact, asked me to explain to her class how Thonglaw could impregnate not just Belle but also Rosy and Tuy Hoa. Her young students had been raised to believe in monogamy and couldn't fathom a philandering elephant! (continued on page 70)

Just For The Elephant, Belle Tells Her Story

Implicit in the statement, "An elephant never forgets," is that an elephant thinks. But what does an elephant think? Particularly, what does Belle think? Here is the answer.

By BELLE, THE ELEPHANT
as told to
Mrs. VIOLET L. SHAFER
Eugene Housewife

People say that, "An Elephant Never Forgets." That is the understatement of the year! I for one, will never forget all the confusion that is going on in my home, The Portland Zoo. You'd think that no one ever had a baby before. Well, come to think of it, I don't recall hearing about any elephant babies being born around here recently.

I suppose that's why all the people are hanging around here all the time. Taking pictures, Holy Bologna! I never had so many pictures taken in all my life. They look at me as if I were some kind of model. Me! with a thirteen and a half foot waist line!

It won't be long, till my little offspring will be born. Never being a mother before, I'm getting pretty nervous. But if you think I'm nervous, you should see my master! He's a WRECK! He and his reporter friends have moved into my home. They even have the gall, to sleep on my hay. Now hay is something I like fresh, not all tainted up with people smell!

I've tried to entertain my guests, as best I could. I even learned to do the twist. No! Boy! do I get attention when I do that! They start right in walking the floor with me. I guess they are afraid I'll go into orbit. I'll say this, when I get all of me in motion, it is a little hard to get me stopped.

I have three lady friends living with me here. Housing is an awful problem. Two of the ladies are expecting, too, but so far, I'm getting most of the attention. Our Papa lives with us ordinarily, but he was moved out, when the reporters moved in. He sure is unhappy, and every once in a while, he really lets the people know how he feels. He goes into an act, somewhat like a People Temper Tantrum, only Bigger and Badder!

Now you may think this is an elephant's tale, but I've been having the miseries sumpin awful here lately. I get so moody sometimes. I'd just love to fill up my long trunk, and squirt a whole tubful of water in some unsuspecting reporter's face, or pick him up with my trunk, and dunk him good! for sleeping on my hay! Boy! That really bugs me!

Summing it all up:
I get the best of food each day,
Carrots, apples, dirt, and hay.
I guess I really shouldn't complain,
With a home like mine, out of the rain.
Someday they'll write a story too,
About the Elephant, born at the Zoo.
This may all sound a little wacky to you.
But this is all written, FROM THE ELEPHANT'S POINT OF VIEW!

4120 Marshall Ave.
Eugene, Oregon

This cartoon, published in The Oregonian *on February 1, 1962, illustrates this letter written by a Eugene housewife—from Belle's perspective. I'm the guy with his head in his hands, and Zoo Director Jack Marks is the cigar smoker. The media is passed out on the hay. (Reprinted with permission from* The Oregonian*)*

Roses For Belle

BABY SHOWER for Belle, pregnant pachyderm, was initiated at zoo Friday by Wayne Herrin, left, of Flowers, Inc., who presents pair of blue booties to Belle (right) with aid of Dr. M. Maberry, zoo doctor. Even Rosy spit out roses, ate stems. Director Jack Marks encourages shower, says "Bring Money, Keep Elephants Green."

World Kept Waiting For Elephant's Birth

Belle, the belle of Portland's zoo, Friday was still keeping the world waiting for the birth of her first child, the first elephant to be born in the United States in 43 years, the first ever conceived in a U.S. zoo.

Belle pushed the panic button at the zoo Thursday morning when she began to feel inexplicable pains in her ample abdomen. Dr. Mathew Maberry, zoo doctor, who had been keeping a 24-hour watch on the expectant mother, cleared the decks for action. Morgan Berry, who owns Belle, Thonglaw and Pet, rushed down from Seattle.

Thonglaw, the busy bull who fathered Belle's baby and two others in the past 18 months, was lured into the waiting room, where he began pacing the floor in the manner expected of expectant fathers.

Belle remained in the main "maternity ward," attended by three other females, two of which are also pregnant — Rosy and Tuy Hoa. While the public can see the expectant elephants through the glass of the elephant house, Director Jack Marks has banned all flash bulbs and strobe lights, which might "spook" the elephants, and will close the house to the public if the attending physician feels it necessary, he said.

Labor Pains Noted

Belle experienced labor pains at intervals ranging from a half hour to two hours through the long night. Dr. Maberry's examination showed tendons in the pelvic arch fully relaxed. Her body temperature dropped slightly. The baby—or babies—had dropped into position for delivery and could be seen kicking and squirming occasionally.

But the petulant pachyderm, like many another mother, declined to deliver. Dr. Maberry, who slept in the elephant house Thursday night, is inclined to think the blessed event is most likely to occur in the still hours of the morning, judging by meager records of the four or five other elephants born in the U.S.—none of which lived a year.

He is inclined to feel even elephants like privacy at such sacred moments and Director Marks has limited the press to one still and one movie photographer for the big event—plus one Life photographer who flew in at midnight for the historic birth.

Berry believes the baby will weigh 200 pounds—or it could be twins, a rare, but not unknown event in elephant herds.

Every day, it seemed, someone from the community came to the zoo to present the expectant mother with a special gift, such as giant booties stuffed with roses. Belle's pregnancy truly was a collective experience for all of Oregon. (Reprinted with permission from The Oregonian*)*

Belle often received gifts in her maternity ward, and I was usually on hand to help present them. (Reprinted with permission from The Oregonian*)*

I must admit, I felt a twinge of sympathy for the media; waiting for an elephant to give birth is mostly a boring—and smelly—job. An elephant poops on average six times a day, dropping a total of about eighty pounds of dung. Between Belle, Thonglaw, Rosy, Tuy Hoa, and Pet, the elephant barn filled with four hundred pounds of poop daily. Elephant keepers scrambled to clean it all up, but the elephant barn consistently stunk like rotting potatoes. That meant those of us on the Great Portland Elephant Watch stunk, too. *Oregonian* reporter Lev Richards wrote that his wife recoiled every time he raced home for a shower and a clothes change. And *LIFE* magazine writer Shana Alexander said her plush Portland hotel, The Benson, gave her a room with two bathrooms because she smelled so bad.

Zoo's Elephant In Labor But Reporter Gets Pains

Editors's Note: Leverett Richards, Oregonian Aviation Editor, has been assigned to the maternity ward at the Portland Zoo to await the birth of Belle's baby.

DEAR BOSS:

I got troubles.

My saintly wife is very understanding, very.

I was away once for 14 months. Just told her I had to go to the North Pole. Next time I was away for nine months. Explained I had to go to the South Pole.

But I am gone for three days and nights and she gets all excited.

I explain I'm sitting up with a sick elephant friend and she takes off like an astronaut from Cape Canaveral.

You tell her! You sent me to the zoo to help Belle have her baby. Said it would just be for the night. (That was way last Thursday.)

Looked at my proboscis — remember? — and said: "You look more like an elephant than anyone around here. You just volunteered."

So I sleeps with these elephants for three nights — make that four come Sunday — and nothing happens.

Proof Lacking

"You said you were having a baby. Prove it," my wife says on the phone. She sounds cool.

So I gets maternity leave and dashes home for a half-hour reunion and a change of undies.

My loving wife greets me with open arms — and an open bottle of Airwick.

"I need a nap and a bath," I announces heartily.

"You need delousing," she insists. "And leave your clothes outside. You smell like a barn."

"I am a barn," I explains, in some confusion. After all, seven days without sleep makes o n e weak. "That's where I'm living. What do you expect, Chanel No. 5? . . . And if my clothes look like I'd slept in them, that's because I did."

"You never sat up with me three days and nights when I was having a baby," she reminds me, with that saintly calm which is worse than the storm.

"I know, darling, but you weren't having an elephant. If you had an elephant, now, we'd ALL sit up with you— television, the press and Life magazine. That's a promise."

How generous can a guy get—and in full color, yet.

"What do you see in this Belle creature," she inquires coldly. "She's shaped like a bag of sauerkraut. Ought to try Metrecal. Really, dear, she's ridiculous. Why, elephants aren't even people.

"Looks like she had been designed by a committee and the committee never reached a decision. A living monument to a committee, that's what she is. Made out of old spare parts. And they don't match.

"How can you fall in love with an elephant? You can't even tell which end is which."

Now I ask you, Boss, how do you explain elephants to a woman who thinks she has been scorned?

I leave it to you.

Your Redolent Reporter,
Leverett G. (Elephant Boy) Richards

As this cartoon shows, waiting for Belle to deliver was smelly business for the media and all other members of the Great Portland Elephant Watch. (Reprinted with permission from The Oregonian*)*

Belle's big pills made handy poker chips for the media, who had hours to kill while waiting for Packy's arrival. Here, Jack Vander White, left, a photographer for the Portland Zoo Society, and Bob Snethen, a cameraman hired to shoot a scientific movie for Morgan Berry, played a hand or two in the elephant barn's feed room. (Reprinted with permission from The Oregonian*)*

Reporters and photographers passed the time playing poker, using Belle's large liver pills as poker chips. And they drank endless cups of coffee, some of which they shared with Belle and the other elephants, who not only swigged the coffee, they munched on the paper cups. The media often lounged on a hay pile, with a view of Belle, and swapped elephant stories. Like me, they read everything they could find on elephant births, which wasn't much.

Finally, desperately bored, the press crew created a secret order called SPOPE: the Society for the Preservation and Observation of Pregnant Elephants. They practiced secret handshakes and grunts, and vowed to continue meeting long after Packy's birth. Somehow they roped me in to their society and eventually elected me Eternal Elephantrician. It was pretty silly stuff; the society's salute, for instance, was a Statue of Liberty pose, with the right arm held high so that an elephant's trunk could "elegantly" sniff the underarm. But it passed the time and entertained the press crew, who couldn't help but bond through the long wait.

All of the media were men except for Shana, who was a particularly big fish in this hometown crowd. In an era when most women worked in newspaper society sections, Shana had blazed her way up through the ranks to become *LIFE*'s first-ever female columnist and writer. Later, in the 1970s, she would go on to even greater fame as the liberal voice on the "Point/Counterpoint" segment of television's *60 Minutes*. In 2000 she authored her book *The Astonishing Elephant*, much inspired by her endless days in Portland's elephant barn. (Shana was kind enough to dedicate the book to me, writing: "For Dr. Matthew Maberry, DVM, my first teacher.") Shana's eleven-page spread in *LIFE* detailed Packy's birth and catapulted the event from a Pacific Northwest phenomenon to a national triumph.

Portlanders, naturally, showed the most excitement about the much-anticipated birth. The zoo's switchboard fielded more than 500 phone calls a day, mostly from folks seeking any and all tidbits about Belle's condition. Some callers, though, wanted inside information about possible birth dates or a predicted baby weight so they could place their bets in one of the endless elephant betting pools. Such pools even included international guesses. The 350 passengers aboard the luxury liner SS *Brasil* of the Moore-McCormack line (which included 12 passengers from Portland) started a pool while on the high seas.

"Prognostications vary from 300 to 640 pounds, trunk and all," the ship's chief purser radioed *The Oregonian*.

Beyond being curious, some callers wanted to help to speed this slow, plodding birth along. *The Oregonian* reported how one caller suggested putting teak logs in the elephant barn to make the birth room seem more like Belle's native Thailand. Another caller thought that Thonglaw should give Belle a "fatherly kick in the butt to get the show on the road." And yet another advocated privacy for Belle: "No mother is going to have her baby while all those doctors are watching. If she did, she wouldn't be a lady."

What else could I do? I just chuckled over the suggestions and continued what I'd been doing.

Waiting.

"No accurate records have been kept in the jungles, and there has been no chance to keep records in zoos. Dr. Maberry will be the first to keep exact, certified records and publish them.

—Morgan Berry, as quoted in *The Oregonian*, 1962

Doctor Naps Awaiting 'Delivery Room' Call

I have to say, a hay pile is a really scratchy mattress. But more often than not, that's where I ended up nights while waiting for Belle to deliver. (Reprinted with permission from The Oregonian*)*

March 1962: No Rest for the Weary

I'm not a man who needs a lot of sleep. In my teens I rose daily at 4 AM (a habit I continue to this day) to do chores before heading to school. At Washington State University my veterinary studies demanded long hours of study; out of 1,200 applicants, just 40 were accepted and only 15 of us actually graduated. Also I was busy with family life. I married my first wife, Hazel, in 1942 and had two children, Sondra and Wake, who both still live in Sequim. Hazel and I divorced shortly after my graduation from veterinary school in 1947, but my schedule hardly slowed down. A new vet on the rise, I opened a practice in Port Angeles, near Sequim, and my territory extended to the Pacific Ocean, a two-hour drive away.

During Belle's pregnancy, it felt natural to continue my routine of four to five hours of sleep a night. But just because Belle was having a

Here I am in with my son, Wake, who still lives in Sequim, Washington, with his wife, Connie. Wake and Connie blessed us with two grandchildren and five great-grandchildren. (Courtesy of Dr. Matthew and Patricia Maberry)

baby didn't mean I could focus solely on her needs. I had a zoo full of other animals that also needed my attention, and I gave it to them on a daily basis.

Typically, I arrived at the zoo around 6:30 AM and always checked first on the animals needing the most care. During Belle's pregnancy, my first stop was the elephant barn. Thonglaw and the girls always seemed happy to see me, often unfurling their trunks to trumpet a greeting. Satisfied that all was well, I then took off walking—briskly—to make my morning rounds of the zoo's 1,000 or so other animals.

I love all animals, but primates—who are most like humans—can be a feisty lot. More than any other animal, the chimpanzees, orangutans, monkeys, and apes goofed around and gave me grief. Sometimes they bit. But they also had their charming side. During one visit, the parents of a brand-new baby greeted me, and the father held up the furry chimp for me to see, showing it off like any proud dad. All that was missing was the cigar in his mouth! Another day, I found a couple of orangutans that had somehow escaped their enclosure and moseyed down to the California sea lion pool. The pair just stood there and watched the sea lions frolicking. When they saw me, one offered a hand, which I took, and we all strolled back to the primate house.

Over the years, I've come up with many methods to improve animal medicine, and the primates inspired one such discovery. When I first started working at the zoo in 1958, many primates suffered from tuberculosis, which stressed their systems and created an electrolyte imbalance. Today, humans can just drink a glass or two of Gatorade to replenish their electrolytes, but no such beverage existed in 1962. Primates became lethargic when lacking enough electrolytes, and some even died. After I figured out the imbalance and corrected it, our primates felt much better. I wrote about my findings, which helped vets worldwide to employ the same cure.

On my morning rounds, though, I was mostly just an observer; you can tell a lot about an animal just by looking at it. What they eat (and poop) also explains their general health. Most of all I remembered that wild animals will always be wild animals—and that you have to be respectful of that.

I always carried candy in my pocket—sometimes Life Savers but really anything I thought the animals might enjoy. The raccoons and beavers especially got excited seeing me, knowing I had treats.
(continued on page 82)

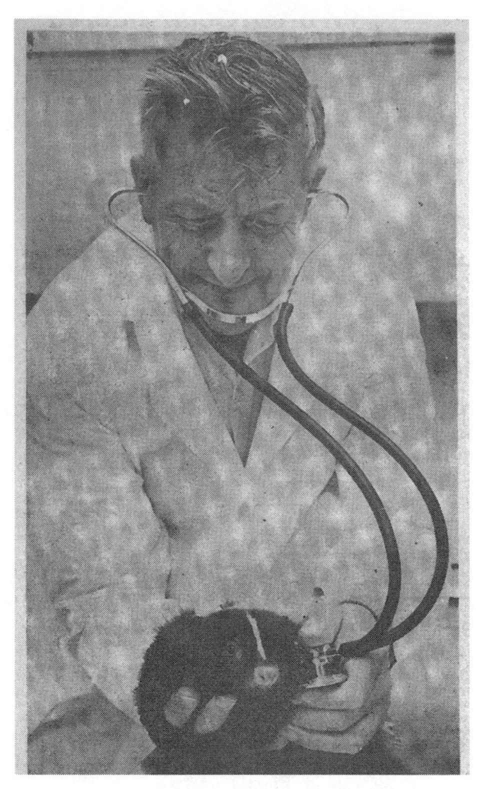

(Left) Belle's pregnancy dominated my attention, but I still had to attend to a zoo full of animals — and sometimes, animals from elsewhere. This seal arrived at the zoo after being found in Gales Creek near Forest Grove, Oregon. (Courtesy of The Hillsboro Argus*)*

(Right) Thank goodness this skunk did not spray me while I checked its heartbeat! (Reprinted with permission from The Oregonian*)*

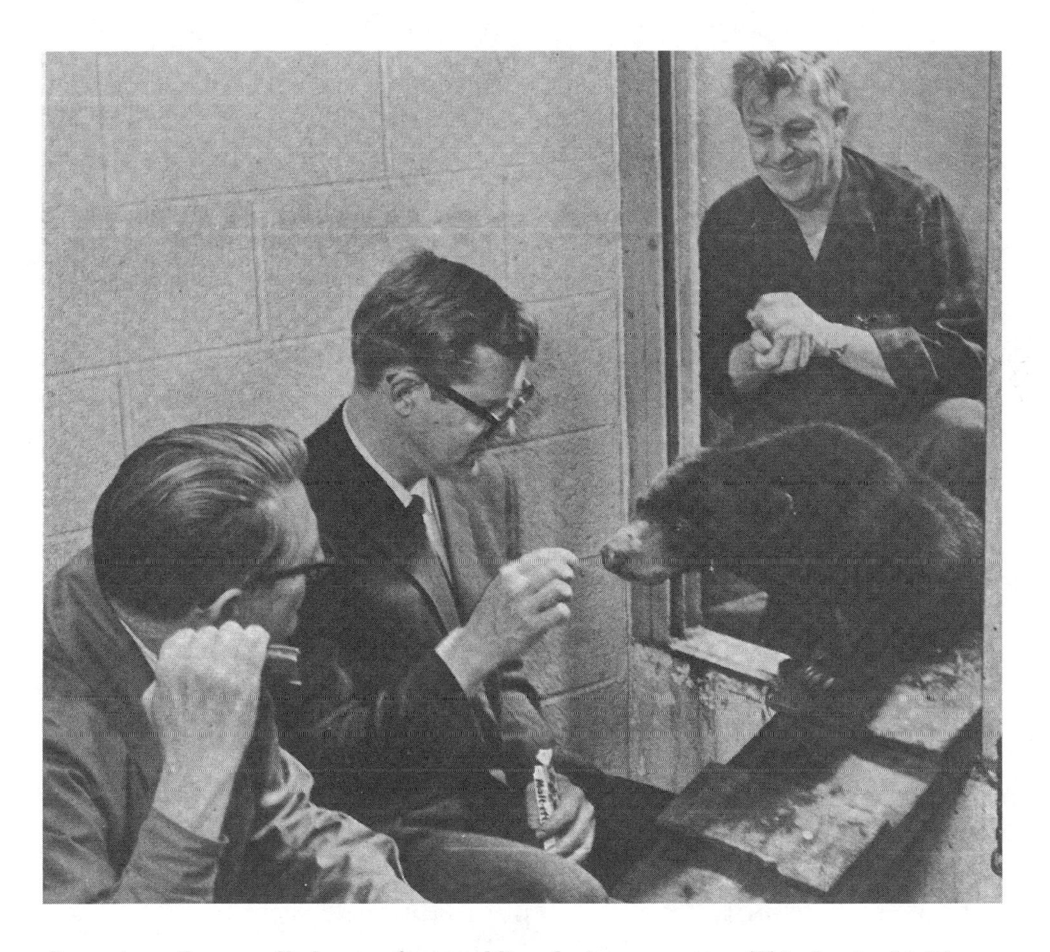

Sometimes I was called upon for a public relations moment. This day in 1967 I took along Boo, a Malayan sun bear who had just arrived at the zoo, to greet Howard S. Mason (left), president of the Portland Zoological Society, and Lawrence Curtis, the society's new executive secretary. (Reprinted with permission from The Oregonian*)*

I wish the bears had been as excited to see me, but bears are moody, which means they're sometimes dangerous. I vividly remember one night at the old zoo when a polar bear somehow got out of its enclosure. Night watchman Lawrence Coppy discovered the escapee and the bear chased after him, intent on making Coppy a meal. Somehow, in between running laps around the bear exhibit, Coppy reached an emergency phone to call Jack Marks. "I really could use a little help up here!" Coppy screamed. Jack called the other keepers and me, and we all raced to help. Jack brought a rifle, but as soon as the bear spotted him with the gun, he took off in his direction. Jack managed to scramble up a ladder leaning against a building but dropped the gun in the process. Finally we opened up the gates to the bear enclosure and coaxed the renegade to chase a keeper inside. The bear ran in, the keeper ran out, the gates closed, and we all exhaled a huge sigh of relief.

Life is never boring at the zoo—even when you're saying good-bye to an animal. We got rid of our lone gator when we moved from the old zoo; unlike today, people back then could buy an exotic animal from the zoo. Jack sold our alligator to two men who claimed they owned other alligators and knew how to handle them. It sure looked that way when they came to get their purchase: they flipped him over, rubbed his tummy, and put that reptile right to sleep. Then they laid the alligator in their car's front seat, between the two of them, said their good-byes, and took off down the road. Well, they'd driven maybe 100 feet when all of a sudden both car doors flew open and the two men jumped out. Apparently the alligator woke up and was snapping mad about its new surroundings. After a good laugh, I got some rope and helped secure the gator for the ride home. I must say, both men looked a little embarrassed as they climbed back in their car and drove away.

This old leather medical bag, packed to the brim with the tools of my trade, has traveled daily with me for so many years I can't imagine practicing medicine without it. (Photo by Michelle Trappen)

During Belle's pregnancy, I always revisited the elephants after my morning rounds, and then I'd take off for lunch. Often I drove fifteen minutes downtown to dine at Henry Thiele Restaurant. The white stucco landmark, at the corner of Burnside Street and Northwest 23rd Avenue, opened in 1932 and had a real old-school feel. Among a host of regulars, I loved the place. I gave daily updates on Belle's condition to my friends there. Sadly, Thiele's closed in 1990 and has been replaced by a bunch of contemporary stores.

After lunch and back at the zoo, I usually performed surgery. But not in a hospital; the zoo didn't have one until 1966. Rather I'd grab

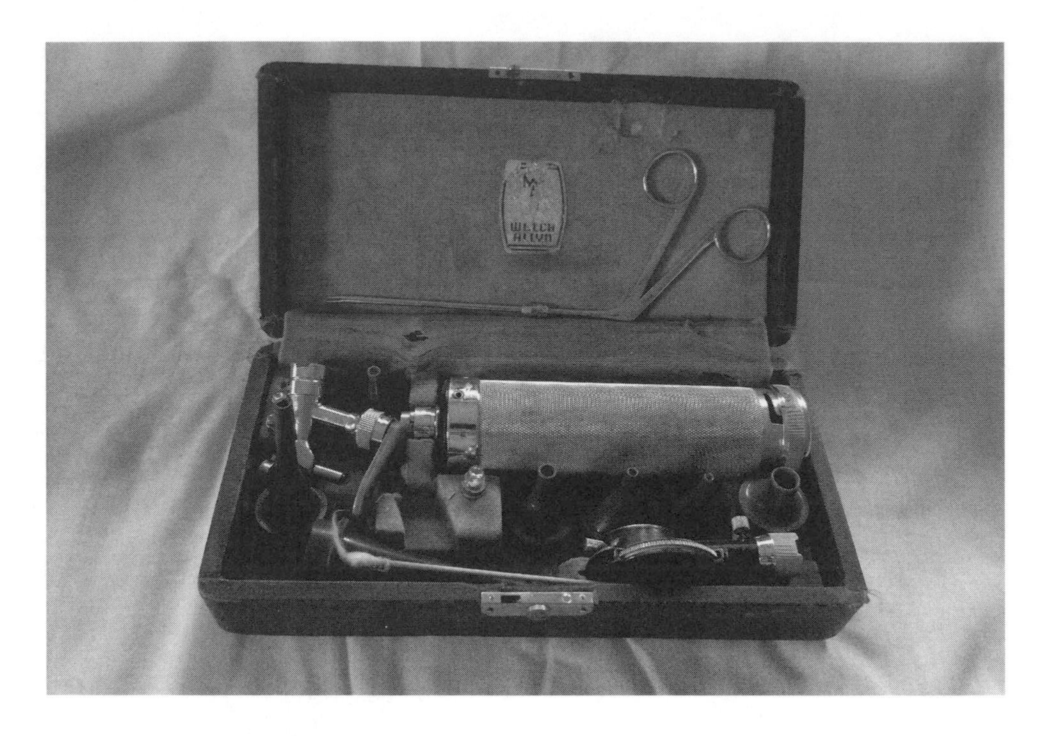

I've used my otoscope to examine the ears of thousands of animals throughout the years. (Photo by Michelle Trappen)

my fifty-pound leather bag—brimming with tools—from the trunk of my Chrysler (I always drove the biggest Chrysler they made) and hauled it wherever I had to operate. Typically that was in the animal's living quarters.

Back then surgeries could be tricky. Anesthesia for animals was just being developed, and it was a guessing game to figure out how much each animal could handle. Often I fell back on my farm skills to get the job done. Elk, for instance, periodically needed their antlers sawed off; otherwise they poked them through their fenced enclosure, creating a safety hazard for zoo visitors. But no elk was going to stand still and let me saw off his rack! So, each time, I recruited a few strong

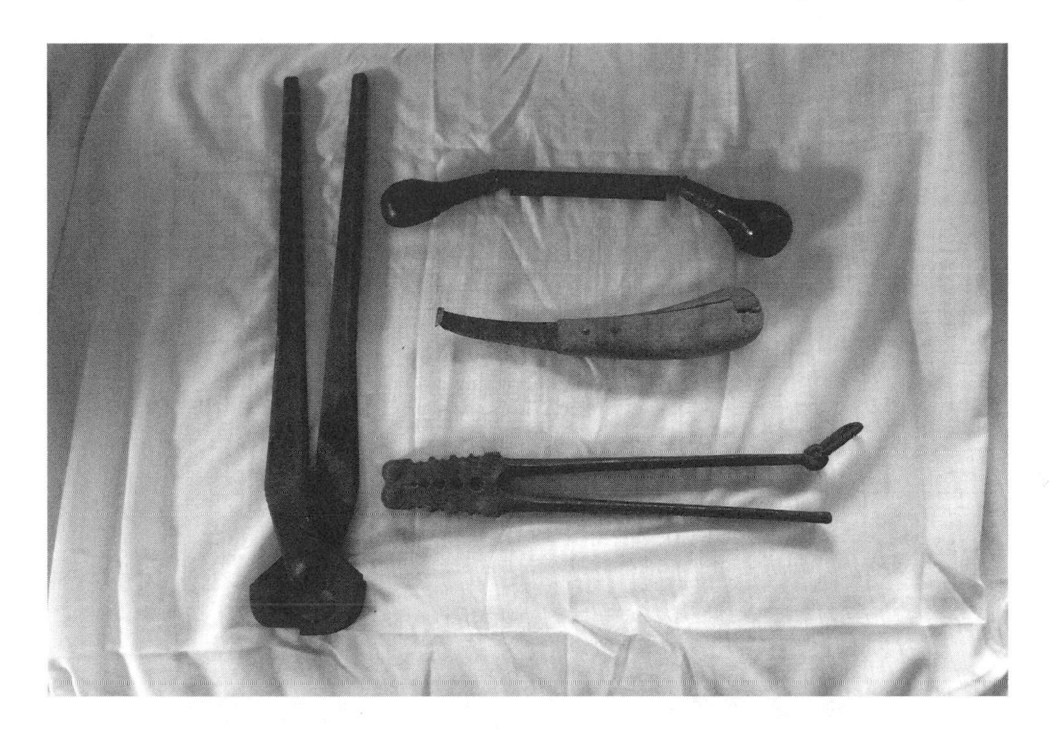

Here is an example of the medical equipment I regularly used. The top right blade, for instance, helped me trim elephant feet. The tools look archaic by today's standards, but they worked fine. (Photo by Michelle Trappen)

handlers to help me lasso the elk's neck and feet. Then, while the handlers pulled the ropes taut, I swiftly sawed off the antlers. I learned to work fast for the animals' sake.

Later in the 1960s, I connected with a physiologist named Tony Harthoorn, who had developed an animal tranquilizer called M99. That drug was a godsend. I could put monkeys under easily and they recuperated really well.

Other animal drugs were in various development stages. I contacted the drug companies and they contacted me, so I could try whatever new drugs were available; in some ways my zoo animals were test subjects.

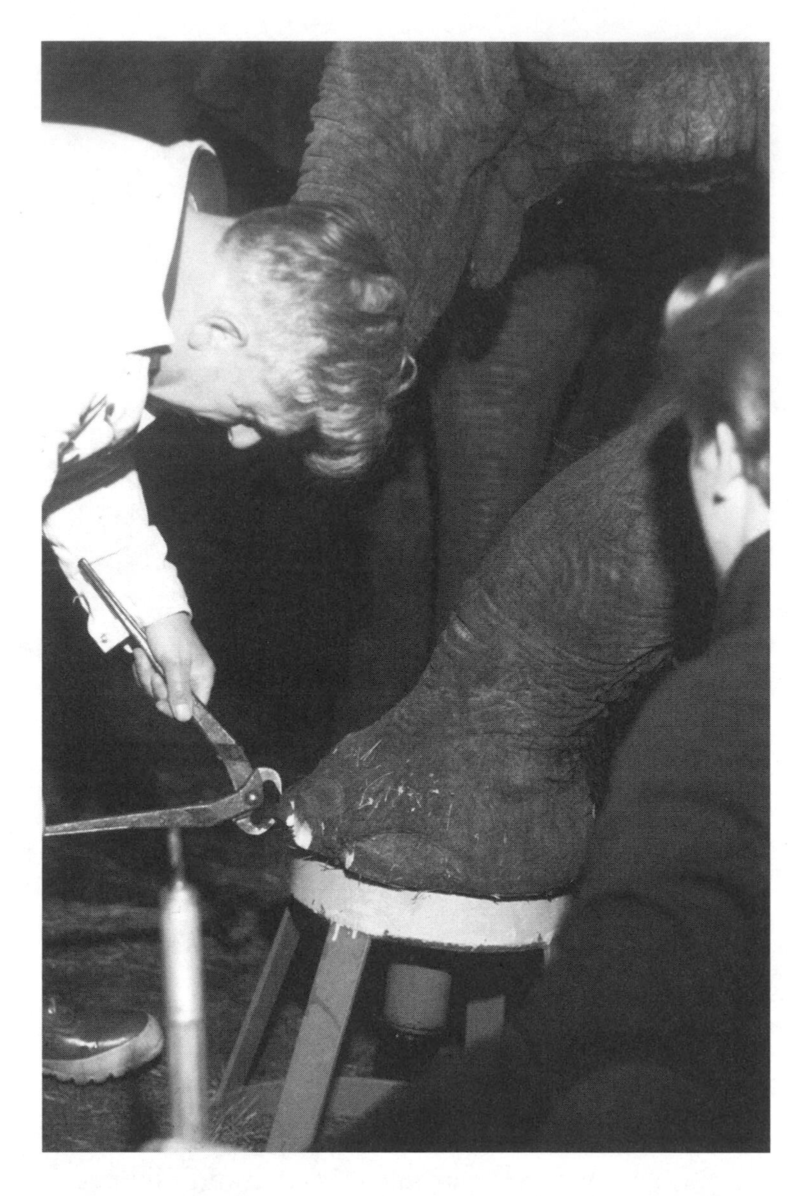

I often trimmed elephant feet to keep them in good shape. Captive elephant feet need regular care since the animals do not have the space to walk like they would in the wild—an activity that naturally maintains their feet. (Courtesy of Jim Anderson)

Looking back, it was all rather controversial because you never knew just how much of each drug an animal could tolerate. So I started small and adjusted accordingly. It was a risk I was willing to take since I was desperate to find drugs that minimized animal suffering.

Later in the 1960s, when Packy was an adolescent, Dr. Marlowe Dittebrandt worked in the zoo's hospital as a pathologist. Ditty, as everyone called her, was actually a physician, but she loved animals. Her expertise helped immensely, and the two of us became close friends. Ditty also became friends with Morgan Berry, and eventually she moved to Elephant Mountain to help with his animals.

Animal ailments often took up my entire afternoon. After that, I would stop by the zoo's kitchen to make sure that the animals' meals were being prepared properly. While pregnant, Belle's daily diet included one and a half bales of the best timothy hay, three gallons of oats, six pounds of bread, fifty pounds of carrots, half a box of apples, twenty-five pounds of bananas, a quarter box of oranges, and ten pounds of freshly dug dirt. (Dirt helps elephants digest their food.)

Finally, I'd end my workday writing up notes about the animals. Before leaving, at least during the Great Portland Elephant Watch, I once again visited the elephant barn.

But my day didn't end there. Beyond my daily zoo duties, I still maintained my private practice, doctoring small and large animals throughout Portland. And late evenings, even during Belle's pregnancy, I often met friends and colleagues for dinner and drinks to conduct "think tank" conversations about Portland and the world beyond.

By the time I got home and crashed, it was typically way past midnight, and my alarm clock was set to ring as usual at 4 AM.

Waiting for Belle to give birth, though, nothing was normal, including my sleep habits. I tried to make it home (continued on page 91)

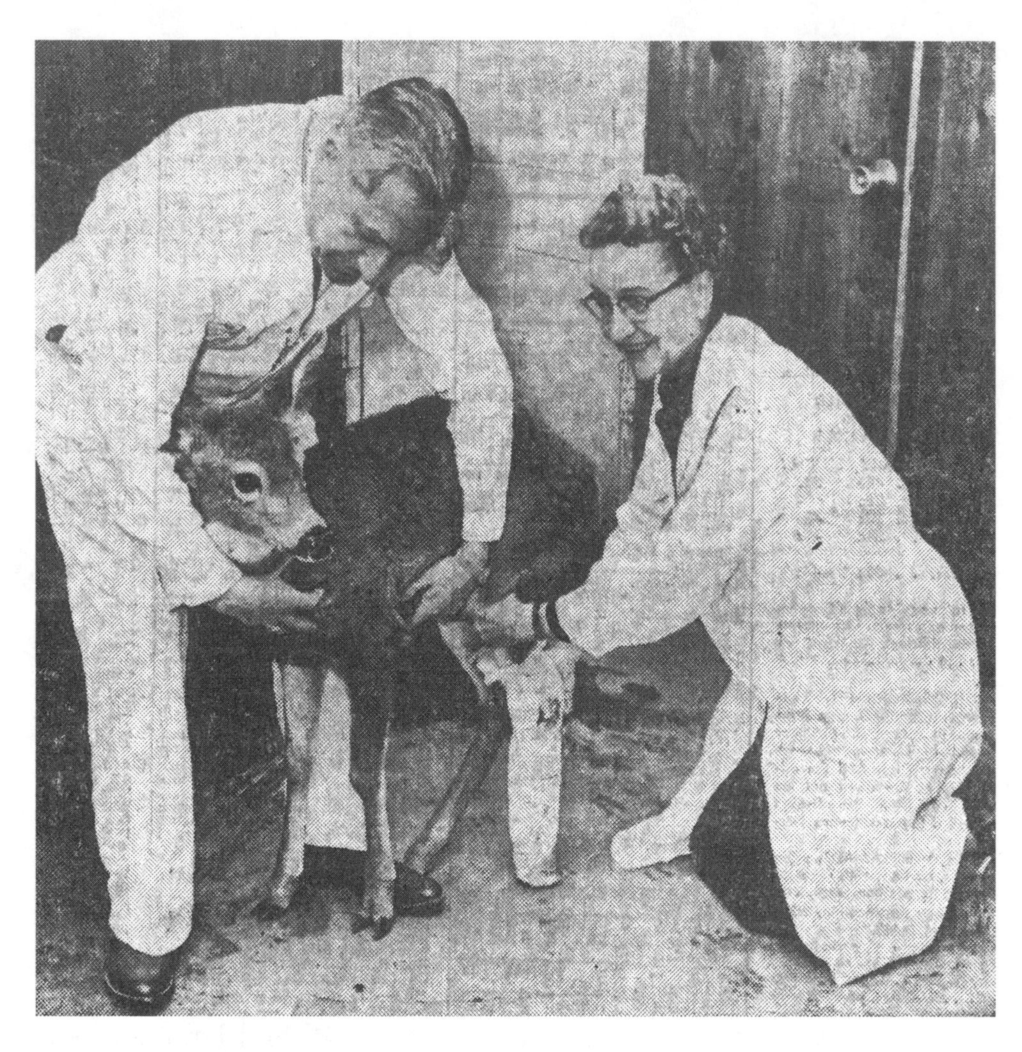

In this picture Dr. Marlowe Dittebrandt, whom I called "Ditty," helped me check the cast on a deer's leg. Ditty was a Portland pathologist who literally walked out of her laboratory and volunteered to work at the zoo. "He [Doc] invited me down to look around," Ditty told The Oregonian *in 1967. "All I could see was an empty room, with a few dirty test tubes and a centrifuge that didn't work. Here was a fine doctor trying to do the work of a whole staff without a helping hand of any kind . . . so I offered to help him." (Reprinted with permission from* The Oregonian*)*

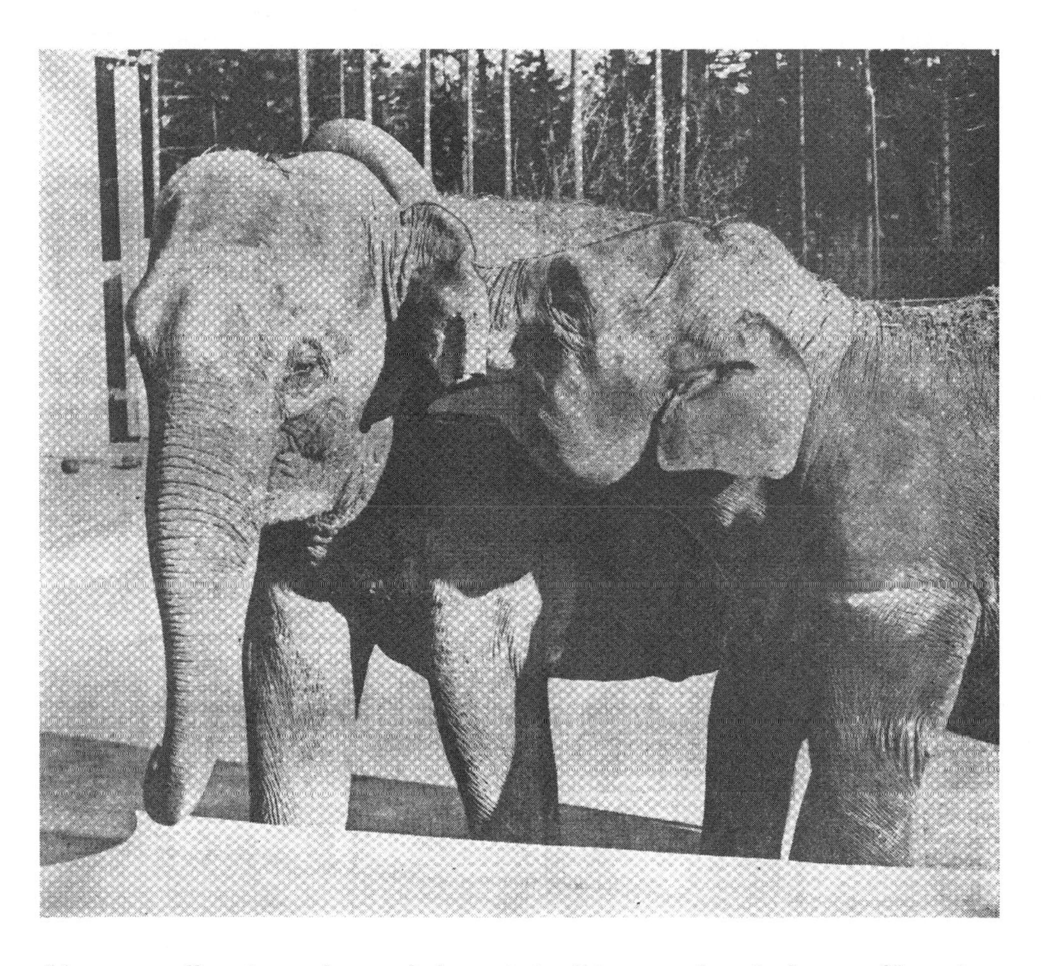

"Are you really going to have a baby, or is it all just another elephantasy?" read the caption that accompanied this photograph in The Oregonian. *Pet appeared to be whispering in Belle's ear the question all of Oregon wanted answered. The caption also reported that most observers guessed the birth date would be Easter Sunday, 1962, which in the end was just one day off. (Reprinted with permission from* The Oregonian*)*

The public loved to involve pregnant Belle in their causes. During the annual March of Dimes "Marching Mothers" drive in 1962, I was given a large tag that read "Tonight I Am a Mother" to present to Belle. Before I did so, however, the tag was changed to read, "Tonight I Hope To Be a Mother," to reflect Belle's drawn-out pregnancy. (Reprinted with permission from The Oregonian*)*

every day, at least to shower and change clothes. But I knew Belle's false labor would one day become real labor. So after finishing a full day's work at the zoo, tending pets in local homes, and having dinner with colleagues, I most times steered my Chrysler back to the zoo.

For a scratchy night of sleep on the hay pile.

Belle's Baby Arrives!

(Reprinted with permission from The Oregonian*)*

April 14, 1962: Fuzzy Face Makes His Debut!

On a chilly Saturday morning in April 1962, Belle finally went into labor. And guess what?

I wasn't there.

Neither was the press.

The night before, a poisoned poodle on the east side of Portland required my attention. Afterward, since I was nearer my Northeast Portland home than the zoo, I opted for my own soft bed over a hay pile. At 12:30 AM I called a zoo guard who reported, "Everything is quiet. Don't worry. Go to sleep."

But at 2:30 AM my phone was jangling and an agitated night attendant, Dick Getchell, was telling me to come.

Quick!

When I got to the elephant barn around 3 AM, I understood Dick's panic. Belle now thrashed about as never before. Hysterical eyes bugged from her sockets as she rammed her huge body against walls and bars, bawling in pain. Maniacally she whipped her trunk up and down, blasting water up and over her head and trembling body. Rosy, Tuy Hoa, and Pet, all at Belle's side and equally frantic, shrieked right along with her.

For two and a half hours, elephant keeper Al Tucker, Jack Marks, and I watched Belle's utter agony from the safe side of the bars walling the twenty-foot by eighty-foot elephant enclosure. I didn't want to interfere or risk being injured (who knew how elephants acted during childbirth), but I wanted to be close enough to dash in if necessary.

Belle's unremitting pain, constant wall bashing, and otherwise odd behavior convinced us that delivery—finally—was imminent. Jack called Morgan in Seattle and told him to hurry down, but Morgan, who had rushed to Portland several times before, said "oh yeah" and went back to sleep.

Around 5:30 AM Belle quieted down and made me wonder if this again was false labor. I even lay down on the hay outside Belle's quarters, anxious for a couple of winks.

I'd barely shut my eyes when zoo foreman Bill Scott hollered that the baby was on its way. Belle was now fanatically crossing and uncrossing her back legs. Then, at 5:56, Belle started spinning like a merry-go-round gone mad. Two minutes later, at 5:58 AM, Oregon's most anticipated baby *ever* slid from Belle, back-feet-first, onto the straw-cushioned barn floor.

The umbilical cord broke as the baby fell to the ground, and Belle made no effort to twist the cord or tie it in a knot. Instead, using her trunk and gently kicking the newborn in the rear, the mother nudged him to his feet. Wobbly, Packy took his first few steps.

The sight entranced the few of us in attendance, but it totally baffled

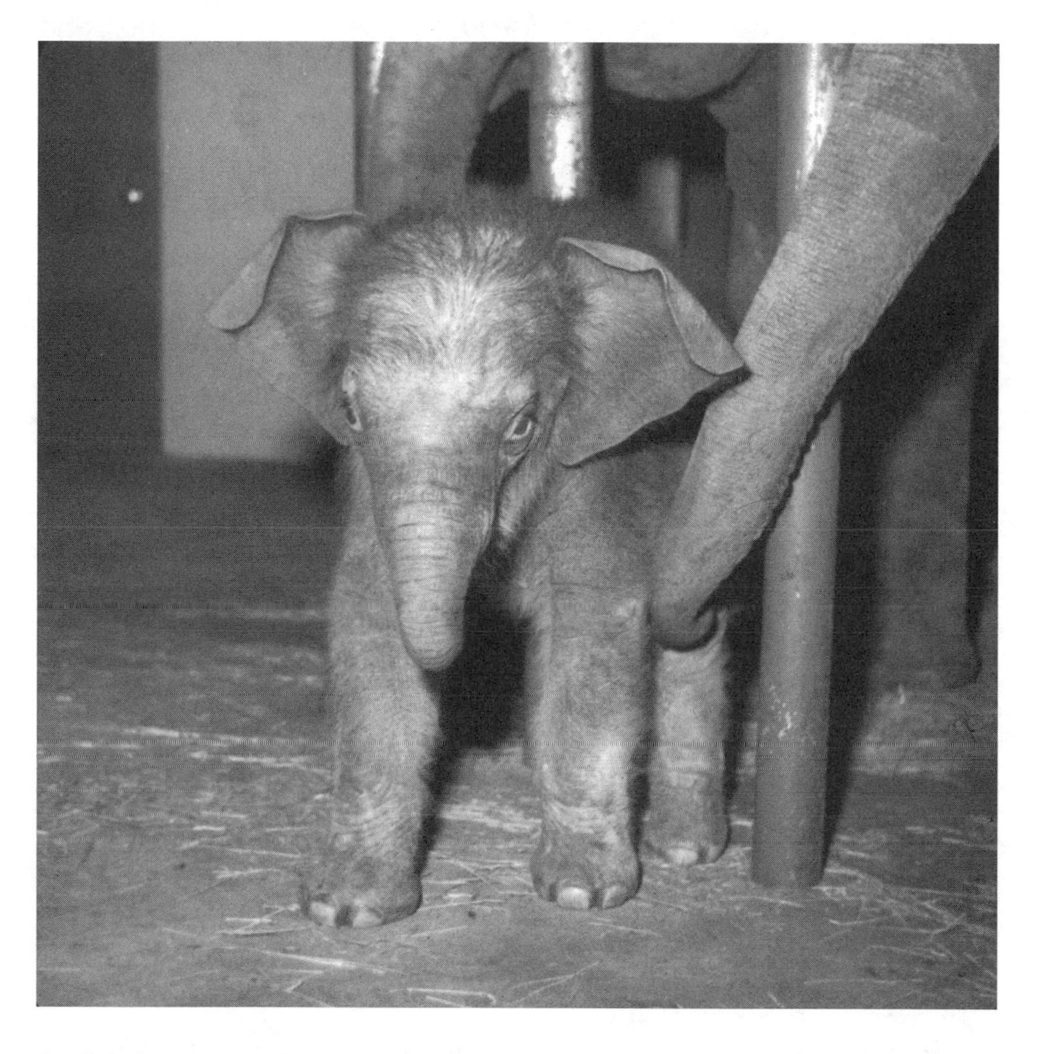

Packy's first few days were all about eating and learning to walk. Here, Belle used her long trunk to help lift Packy to his feet. Packy's hardest early lesson: that dangly, utterly confusing trunk. (Reprinted with permission from The Oregonian*)*

Rosy, Pet, and Tuy Hoa, who had never seen such a thing. Had the trio stayed with their herds in the wild, they would have learned that an "auntie" always helps a new mother. The three females, however, all pregnant themselves, were so feverishly worked up after Belle's endless bellowing that they did what any good friend would do: they tried to sit on this slimy foreign creature on the floor! That's when Al, Jack, and I darted into the birthing room and herded the wailing, novice aunties into another room.

What came next, I have to say, made all the waiting, worrying, and handwringing absolutely worthwhile. Chained circus elephants often tried killing their babies, but not chain-free Belle. Rather, at 6:30 AM she used her trunk to guide Packy's mouth to her breast. The little one must have liked what he tasted because Packy actually squeaked in delight.

In the other room, perhaps realizing this day's triumph, Rosy, Tuy Hoa, and Pet trumpeted loudly.

Jack immediately called Morgan, who raced down to Portland. Then Jack, on the phone with reporters by 7 AM, got so worked up he passed out and had to be rushed by ambulance to Good Samaritan Hospital. Doctors determined Jack strained himself trying to push Rosy, Tuy Hoa, and Pet out of the birthing room, and recommended bed rest. But Jack was *way* too excited. En route home, Jack just had to stop by the elephant barn to get another good look at the long-awaited Packy.

Morgan measured Packy and announced that he stood thirty-five inches high and stretched forty-six inches around the chest. Packy also had a tiny eight-inch trunk, the use of which baffled the youngster at first. The next day, Easter Sunday, six of us helped the baby onto a shipping scale for his first weigh-in. I guessed his weight at between 150 and 175 pounds, but Packy surprised us all, topping the scales at a robust 225 pounds. (continued on page 100)

Elephant Owner Misses Long-Awaited Zoo Event

"When Jack Marks called me in the middle of the night and told me Belle was in labor again, I said 'oh yeah,'

rolled over and went back to sleep," Morgan Berry, of Seattle, Belle's owner and trainer, admitted Saturday.

When Marks called about 6:15 a.m. Saturday to say the baby was born, Berry and his son Kenneth, jumped in their car at Seattle and were in Portland by 11 a.m.

Belle, who was raised in the basement of Berry's home with Kenneth, greeted the two like one of the family, feeling them with her trunk as she does Fuzzy-Face, her bumptious baby.

"Gosh, what big ears."

That was Berry's first comment. Belle has small ears for an elephant, but Fuzzy-Face has ears big even for an elephant, Berry opined. They are bluish purple and fuzzy as a baby robin's breast.

Fuzzy-Face retreated under his mother's belly when Berry sought to get acquainted, but did come out for measurements.

Just Yard High

He stands 35 inches at the shoulder, 36 at the height of the back. His tiny trunk is just 8 inches long — 14 if measured to the base of the skull. His outrageous tail is 17 inches long, with two mandarin mustaches at the end, like a wornout fireplace brush.

He is 46 inches around the chest, 53 around the abdomen, 19 around the front ankle.

Fuzzy-Face became bored with the continuous round of poking, prying picture-taking and measuring and dozed off on his feet. Whereupon Berry tucked mother and baby away in an inner bedroom at the zoo for a much-needed nap.

The new arrival will be weighed for the first time Sunday, Berry said. An official name has not yet been chosen.

Dr. Matthew Maberry first powdered a scratch on the baby's tender trunk with antibiotic powder.

The elephantrician, now the most experienced in the business, almost missed the actual birth.

He was called to the zoo about 2:30 a.m. when Dick Getchell, night attendant, noticed Belle in labor. By the time Dr. Maberry arrived, the baby was beginning to be visible as a bulge in his mother's loose skin, which hangs like a pair of baggy pants around her hind quarters.

Sleep Interrupted

But Belle quieted down about 5:30 and there was no sign of progress, so Dr. Maberry, who has spent many a long, sleepless night in the elephant barn, lay down on the hay outside Belle's door.

No sooner had he hit the hay than Bill Scott, zoo foreman, who was watching the elephants from the top of the delivery room partition, hollered that the baby was on its way. Dr. Maberry got there just barely in time for the moment he had been waiting for since Belle showed her first signs of labor Jan. 18.

The umbilical cord broke as the baby fell to the ground. Belle made no effort to twist the cord or tie it in a knot, as elephant "experts" had said she would do.

Belle's owner, Morgan Berry, missed the big event because — after responding to multiple false alarms — he ignored the call telling him to come quick in the early morning hours of April 14, 1962. (Reprinted with permission from The Oregonian*)*

BIRTH FLOORS DIRECTOR
Jack Marks, Portland Zoo director, carried away

(Left) Poor Jack Marks! Packy's birth excited him so much, he literally passed out. Jack, who loved the zoo he helped create, knew this amazing birth was his crowning achievement. (Reprinted with permission from The Oregonian*)*

(Bottom) Jack couldn't help lighting up and puffing a stogie to celebrate Packy's arrival. (Reprinted with permission from The Oregonian*)*

JACK MARKS, zoo director, who remained at Belle's side throughout night, smokes big black cigar, like proud papa, as he answeres endless phone calls from scientific world which awaits news with keen interest.

Belle Finally Rests As Owner 'Babysits'

Fuzzy Face finally got a good night's sleep Sunday night.

All Belle's bouncing baby boy needed was an "auntie."

Morgan Berry, who has lived with elephants until he thinks like one, finally figured out Sunday night why Belle had not let her baby lie down since he was born 5:58 a.m. Saturday.

Her jungle instinct told her she couldn't protect him alone. In the jungle she would have had an "auntie," a friendly mother elephant to stand guard while she and her baby dozed.

When Berry lay down on an armful of hay and snuggled up to Belle's bars Sunday night the restless mother finally brought her 225-baby over and let him lean on Berry. He quickly dozed off, while Belle stood guard. Then she woke him for his 1 a.m. feeding.

By about 2 a.m. Belle, confident she had finally found a reliable baby - sitter, lay down for her first nap in more than 36 hours, with her darling Fuzzy - Face corralled between her legs and Berry's sleeping form.

"The baby lay so still I though he was dead, until I reached out and touched him," Berry said.

Some people would be insulted at being taken for an elephant's aunt, but Berry is quietly proud. To him, elephants are people — more human, in fact, that some people.

My wife, Patricia, loves this article about Morgan Berry. In fact, it was this article that convinced her that my story had to be told. Morgan's sensitivity to Belle and Packy, as well as Morgan's gentle disposition, touched Patricia's heart. She's often told me how much Morgan and I had in common, especially our true love of animals. (Reprinted with permission from The Oregonian*)*

As Shana Alexander later noted in the *LIFE* magazine spread on Packy's birth, the relief I felt etched my tired face. Months of constant worry vanished as I watched Belle and Packy lovingly interact.

The rest of Packy's birthday is a blur. The zoo held an Easter egg–rolling contest, drawing hundreds of children who—of course—wanted to see the baby elephant. But everybody, children included, had to wait two days (until Monday) to get their first glimpse.

Packy had no name at this point, as we had yet to determine his sex. Elephants exhibit no external signs of gender other than tusks on males, and Packy's tusks would not emerge for several years. The day after the birth, Morgan performed an internal examination that proved the baby was a boy. Portland's Lloyd Lions Club announced the news by hoisting a six-foot by nine-foot blue flag above Lloyd Center, the city's first-ever mall, which opened in 1960.

Now we knew the baby's sex but not his name. We called him Fuzzy Face because black fuzz exploded all over his body. Soon, however, Portland radio station KPOJ sponsored a baby-naming contest, inspiring entries like Ding Dong, Belle Boy, and Nogero (Oregon spelled backwards). But it was Gresham teacher Wayne W. French who won a portable stereo for his winning entry, a catchy play on pachyderm.

As for Packy's mama (who had delivered early Saturday morning), she still had not slept by late Easter Sunday.

In hindsight, we never should have separated the other female elephants from Belle. Without an auntie to guard her newborn, the mother refused to lie down and sleep. Realizing this, Morgan stretched out on an armful of hay and snuggled up to Belle's bars. Then, as Lev Richards wrote in *The Oregonian*, "The restless mother finally brought her 225-pound baby over and let him lean on Berry. [Packy] quickly dozed off, while Belle stood guard. Then she woke him for his 1 AM feeding. By about 2 AM, Belle, confident she (continued on page 104)

I manned the scale during Packy's official weigh-in, while Morgan Berry see-sawed between keeping the baby in place and keeping worried mama Belle at bay. An unidentified person clutched Packy's right leg, while the media and other zoo officials watched. (Reprinted with permission from The Oregonian*)*

On April 14, 1962—delivery day—Belle and wobbly "Fuzzy Face," as he was then called, briefly appeared in the elephant house's observation room. An extremely proud Morgan Berry, who at this point officially owned Belle and her baby, posed with his elephant family. Reporters described the baby as being mud-colored, sprouting long fuzz, and sporting a miniature moustache and beard. (Reprinted with permission from The Oregonian*)*

Two days after his birth, little Packy lumbered alongside his mother in their first public appearance in the elephant yard. (Courtesy of Marianne Marks)

Visitors crowded the elephant house observation room to get their first look at the new baby. (Courtesy of Marianne Marks)

had finally found a reliable babysitter, lay down for her first nap in more than thirty-six hours, with her darling Fuzzy Face corralled between her legs and Berry's sleeping form."

Morgan's days as auntie were short-lived. Tuesday morning, three days after the birth, we reunited mother and child with Aunts Pet, Rosy, and Tuy Hoa, who all nuzzled the furry newborn. As reported in *The Oregonian*: "All crowded around the infant and smothered him with affection. The baby talk lasted for fifteen minutes." Each auntie then took turns watching over Packy while Belle rested.

Morgan Berry proudly led Belle and Packy into the zoo's elephant yard one week after the famous birth. The Oregonian *reported that nearly 11,000 people visited the zoo that day—so many that gates had to remain open an extra half hour. (Reprinted with permission from* The Oregonian*)*

Packy, center, grazed in the elephant yard surrounded by his father, mother, and two aunties. (Courtesy of Marianne Marks)

As for me, the night of Packy's birth I finally went home to my own bed. I hadn't slept in two days. Gosh, that bed felt good—almost as good as my head and heart. Aside from feeling enormous pride, I mostly felt relief. Really, in the end, I hadn't done anything but watch, just like my father had watched countless calves being born.

Nature took its course like I always hoped it would.

Zoo Director Jack Marks visited with Belle and newborn Packy in the elephant barn shortly after the historic birth. (Reprinted with permission from The Oregonian*)*

This Baby's Top Draw at the Zoo

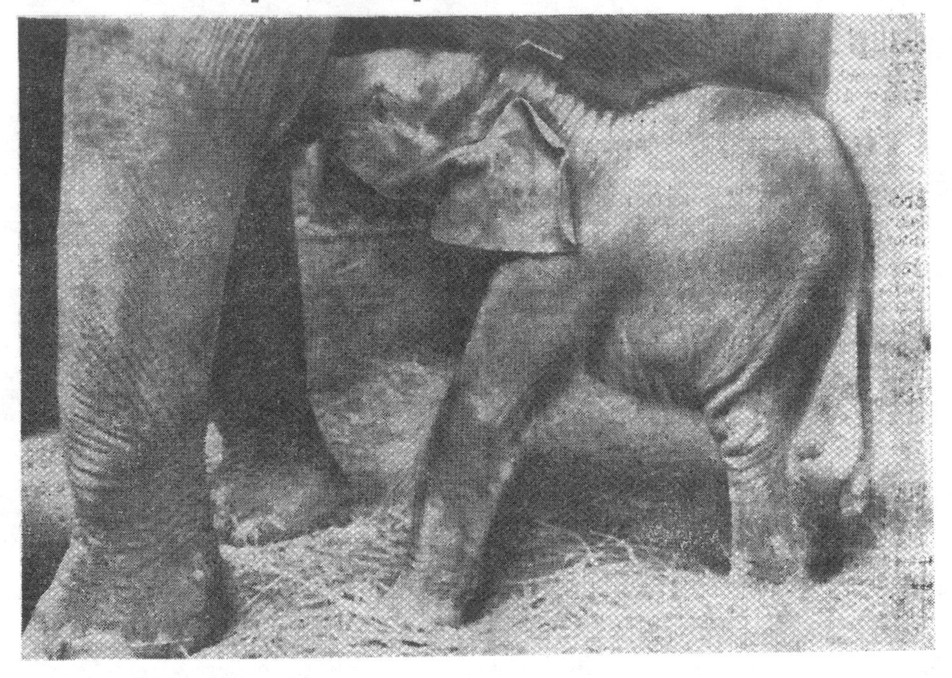

After Packy's birth the public's hunger to see and enjoy him did not dim for some time. Here's Packy nursing at age two weeks. (Reprinted with permission from The Oregonian*)*

Packy Mania

My life blasted off just like an Apollo rocket after Packy's birth.

True, I'd become accustomed to the media hanging around the elephant barn. But only a dozen or so reporters and photographers made up the daily crew that watched Belle's—and my—every move. After Packy's birth, though, reporters from across the United States and overseas called, wrote, or showed up, bombarding me with questions.

Jack Marks got congratulatory "Belle Grams" from officials worldwide, including one from Marlin Perkins of Lincoln Park Zoo in Chicago, who later went on to become the iconic host of the popular television show Mutual of Omaha's *Wild Kingdom*. It read: "Congratulations to you and Belle for [the] first baby elephant to be

Local artists composed at least two songs about Packy's birth, including this one, "'Fuzzy-Face'" Packy, by Portland resident Lewis O. Somerville. The lyrics begin: "The whole world was awaiting the great event for Belle to have her baby elephant . . ." (Courtesy of Dr. Matthew and Patricia Maberry)

Packy's image even graced zoo keys during the 1960s. Zoo visitors used the keys to unlock talking boxes at fifty different zoo exhibits to learn more about the animals. (Photo by Michelle Trappen)

born in an American zoo. Will be waiting for news of second and third."

Oregon stores sold anything and everything Packy, from commemorative plates to ashtrays and medallions. Lucille Frazier of St. Helens, Oregon, wrote one of at least two celebratory songs—"Packy, The Baby Elephant!"—which local television personality Heck Harper later performed. "Packy and His Pals," drawn by Jim Shinn, ran daily in the *Portland Reporter*, and in June 1962 the Portland Rose Parade featured a flower-festooned Packy float.

Through all of this, Packy—thankfully—thrived. The baby nursed

Through the years, I received a lot of elephant paraphernalia—including these pillows—from generous patrons of the zoo. Unfortunately the name of the creator of these handmade pillows has been lost over time. (Photo by Michelle Trappen)

with zeal, and mimicked Belle and the aunties in the elephant yard, especially when it came to bathing. Mother and son delighted the public so much that the first few days after Packy's birth, the zoo stayed open later to accommodate the crowds. Zoo attendance in 1962 topped 1.2 million, a record not surpassed until 2001. Even nationally known dignitaries like Aldous Huxley, author of *Brave New World*, stopped by for a peek at the miracle newborn. Huxley's biologist brother, Julian, also in attendance, said of Belle and Packy, "I've never seen a finer pair."

As spectacularly exciting as the first few weeks were, the zoo, along with the city of Portland, immediately realized they owned neither of

Adwoa Dodoo, center, a visitor from Ghana, had never seen an elephant before she visited the zoo in 1962. With her are Joan Labby and Dr. James Metcalfe of the Zoological Society. (Reprinted with permission from The Oregonian*)*

their star attractions. So before Packy even had his official name, a plan quickly hatched to keep the elephants in the Rose City.

Though Morgan had long displayed his elephants during summers at the Woodland Park Zoo near Seattle, he agreed that Portland's zoo was now the best permanent home. Morgan and I had developed a tight bond, and he knew that Belle and Packy would receive excellent care under my watchful eye. He set a price of $30,000 for them both—a steal, really, given that zoos and animal groups nationwide bid more. Louis Goebel, an animal dealer who supplied animals for Hollywood movies, said he'd pay upwards of (continued on page 116)

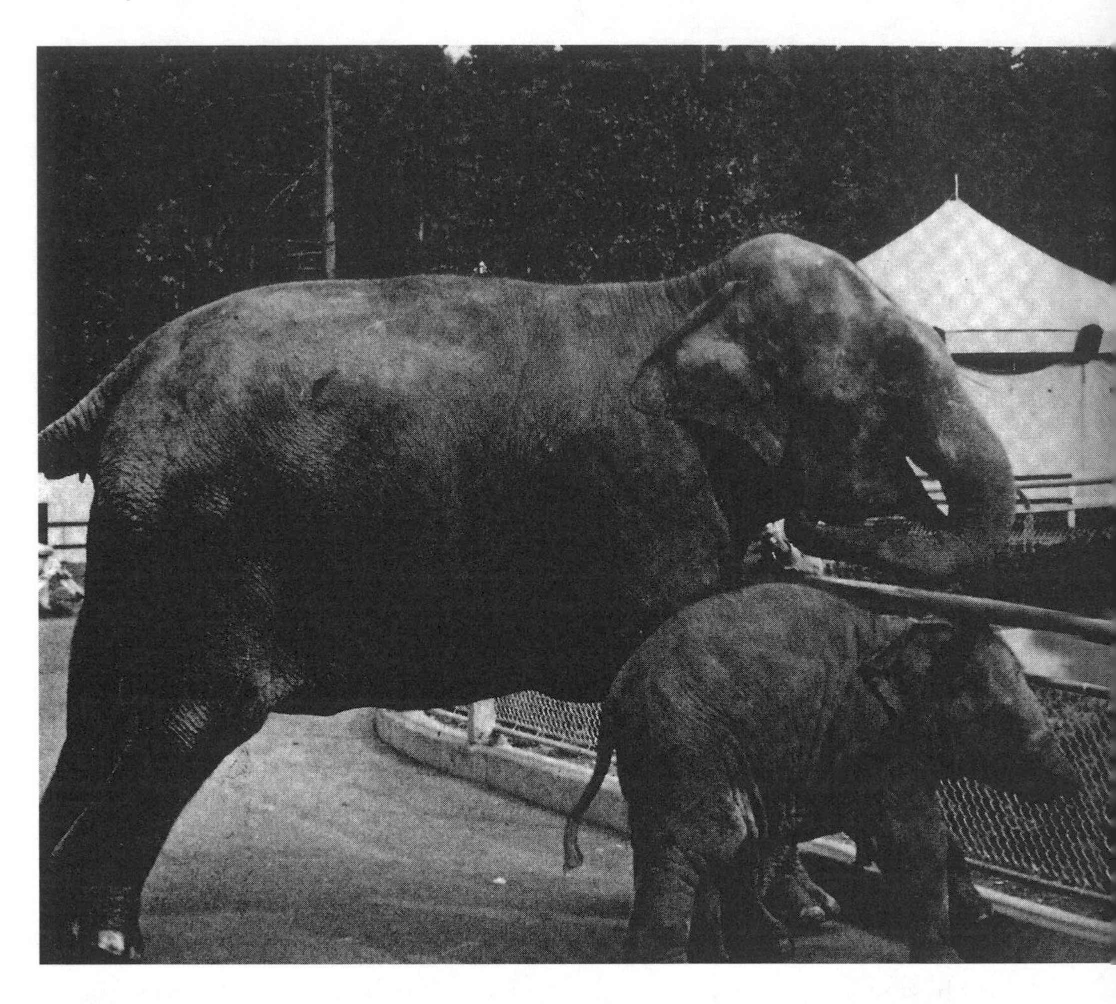

Papa Thonglaw (right), reached out with his trunk to greet Mama Belle and their offspring, Packy, in this rare photograph of father, mother, and baby together. (Courtesy of the Oregon Zoo)

$50,000; a Portland promoter who planned to exhibit the elephants at shopping centers offered $35,000; and the Brookfield Zoo near Chicago expressed great interest in the elephants, knowing full well the huge cost. Morgan, though, resisted all offers, honoring a promise to Jack to give Portland first dibs.

But Portland had to work fast; Morgan allowed just two weeks for officials to figure out the financing. The city's solution: a statewide "Save the Elephants" campaign.

"Belle and her baby have won their way into the hearts of Portland people," Portland Commissioner Ormond Bean told *The Oregonian* three days after Packy's birth. "They are also a valuable asset to the zoo. They will pay for themselves in gate receipts, not to mention the value of the babies. No other zoo in the world can boast a breeding herd of four elephants."

The public, totally enraptured by Packy's birth, rallied around the cause. Organizations mailed in money or ceremoniously handed Jack cash and checks. Others plunked loose change or stuffed dollars into donation cans scattered throughout Oregon. A "Ring the Bell for Belle & Son" rig, operated by the Portland Firefighters Association, rolled through town, allowing contributors to donate and literally ring the bell.

KPTV devoted several hours of television time for a fund-raising drive, a bowling alley sponsored a "Bowl for Belle" moneymaker, and a local car dealer in Gladstone promised to donate $10 from the sale of every new Rambler to Belle and her baby. Fight promoter Vearl Sherman even announced that half the profits from a prize fight in the Portland Public Auditorium (now Keller Auditorium) would benefit the Belle fund.

Schoolchildren also held wildly popular "penny drives" to keep Packy in Portland. Phyllis Kramer's fifth-grade summer school class conducted a typical drive at Parkrose (continued on page 120)

Money Trickles In To Keep Belle, Newborn Baby In Portland's Zoo

Funds continued to trickle into the Portland Zoological Society offices, 4001 SW Canyon Drive, Wednesday to buy Belle and her baby and keep them in Portland, Commissioner Ormond Bean, chairman of the "Jack Marks Elephant Fund," reported Wednesday.

Contributions have not yet been totaled. An additional cashier went on duty Wednesday to help open mail and total contributions, as the "Save the Elephants" campaign got under way.

A "Ring the Belle" campaign is planned, quite literally, with the aid of a bell mounted on a truck, Commissioner Bean said.

He reminded elephant lovers that all contributions to the Jack Marks Elephant Fund are tax deductible if checks are written to the city, the Portland Zoological Society, or the Fund.

Drives Planned

Radio and television stations, some of which had started "elephant - thumping," Monday morning, were each planning campaigns, with all funds being funneled into the special account set up in the city treasurer's office. Marks was named official custodian of the fund for purchase of Belle and her baby, Fuzzy-Face, as yet unnamed officially.

Belle doesn't seem much impressed, but she and her 225-pound darling are creating quite a furore in zoological circles.

Marks is being deluged with "Bellegrams" and congratulatory phone calls from all over the country.

Dr. Ted Reed, formerly of Portland's zoo, now on the staff of the National Zoological Park, Wash., D.C. wired:

"Congratulations on baby elephant. You deserve a record like this. I know of no one nicer it could happen to."

Congratulations Wired

Marlin Perkins, of Lincoln Park Zoo, Chicago, Ill., wired:

"Congratulations to you and Belle for first baby elephant to be born in an American zoo. Will be waiting for news of second and third."

Raymond Gray of Overton Park Zoo, Memphis, Tenn., wired "Congratulations on your zoo's remarkable achievement . . ." So did Arthur R. Watson, director of the Baltimore, Md., Zoo.

Even George W. (Slim) Lewis, hero of Byron Fish's book "Elephant Tramp," wired conditional congratulations:

"If true, congratulations on birth of baby elephant."

Lewis, a former trainer who specialized in "rogue" elephants, had appeared on Seattle radio stations saying Belle couldn't possibly be pregnant.

Visiting animal dealers report Portland is virtually a cinch to win the award offered annually by the Association of Zoos, Parks and Aquariums for the outstanding birth of the year.

The press kept track of various statewide campaigns that fired up Oregonians to keep the mother and her famous baby in Portland's zoo. (Reprinted with permission from The Oregonian*)*

Zoo Director Jack Marks was overjoyed at the donations that poured in to keep Belle and Packy at Portland's zoo. (Reprinted with permission from The Oregonian*)*

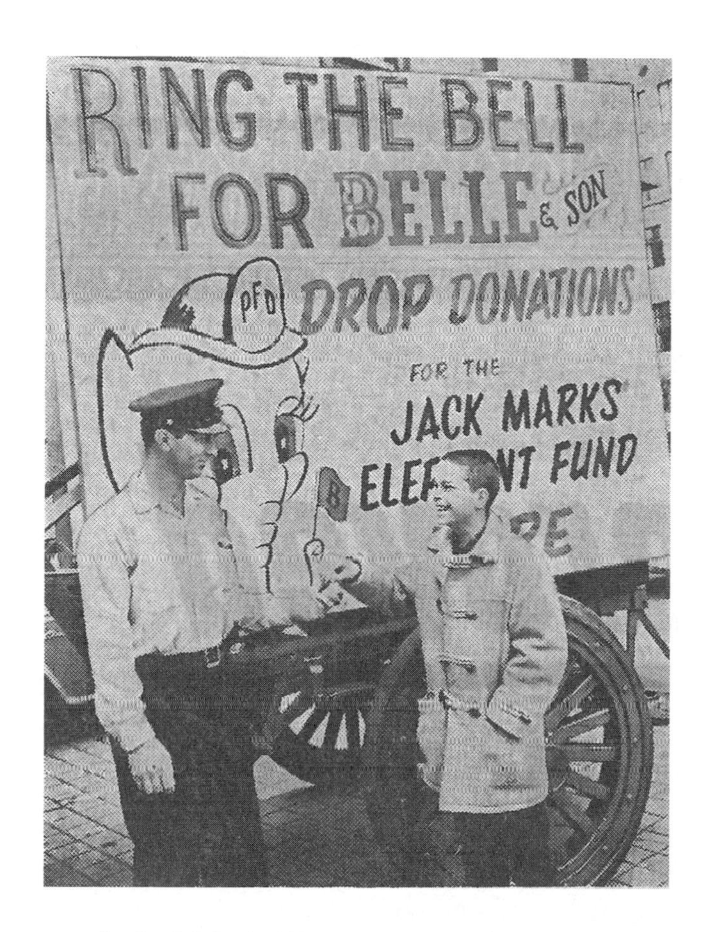

First to donate to the Jack Marks Elephant Fund was Verlyn Grange, twelve, of Cathlamet, Washington. He heard the old fire bell ring as the Portland Firefighters Association's "Ring the Bell for Belle & Son" rig rolled through downtown Portland. The bell is the same one that the fire bureau used to call firemen before the turn of the century. (Reprinted with permission from The Oregonian*)*

> *"Packy's birth was huge news in Oregon, and Doc Maberry was like God. He seemed to be the most important person in the state during that time.*
>
> —Ted Mahar, retired film and theater critic for *The Oregonian*

Elementary: students spontaneously raised $14.75 and Mrs. Kramer matched the sum. In addition, the twenty-nine students spent six weeks stitching Packy a baby blanket.

It wasn't long before Portland had the needed money and the famous elephants forever. Generously, Morgan donated Thonglaw and Pet to keep the tight-knit elephant family intact. "These magnificent gestures by Morgan Berry were as much responsible for the build up of our herd as anything else," Jack told *The Oregonian*.

Suddenly Portland's zoo catapulted from near obscurity to being the preeminent elephant breeding facility in the world. I got calls and letters from zoos around the world that wanted to know how we did it. Zoo officials also came to observe Portland's elephants and to talk to me about the birth process. But it would be years before most zoos attempted breeding elephants; the majority lacked our modern, hydraulic-door facilities, so chaining continued. Up until 1965, when the American Association of Zoo Veterinarians (AAZV) was formed (I was a founding member), communication between vets was difficult: we could talk on the telephone, learn from newspaper articles, or get together. But in-depth veterinarian journals that detailed science-making events like Packy's birth just didn't exist. The AAZV provided veterinarians a forum in which to share the increasing number of innovations in exotic animal care, both in publications and at annual meetings. At one meeting, for instance, I presented my professional paper on "Diagnosis of Pregnancy in the Asiatic Elephant."

All the commotion that followed Packy's birth kept me busier than

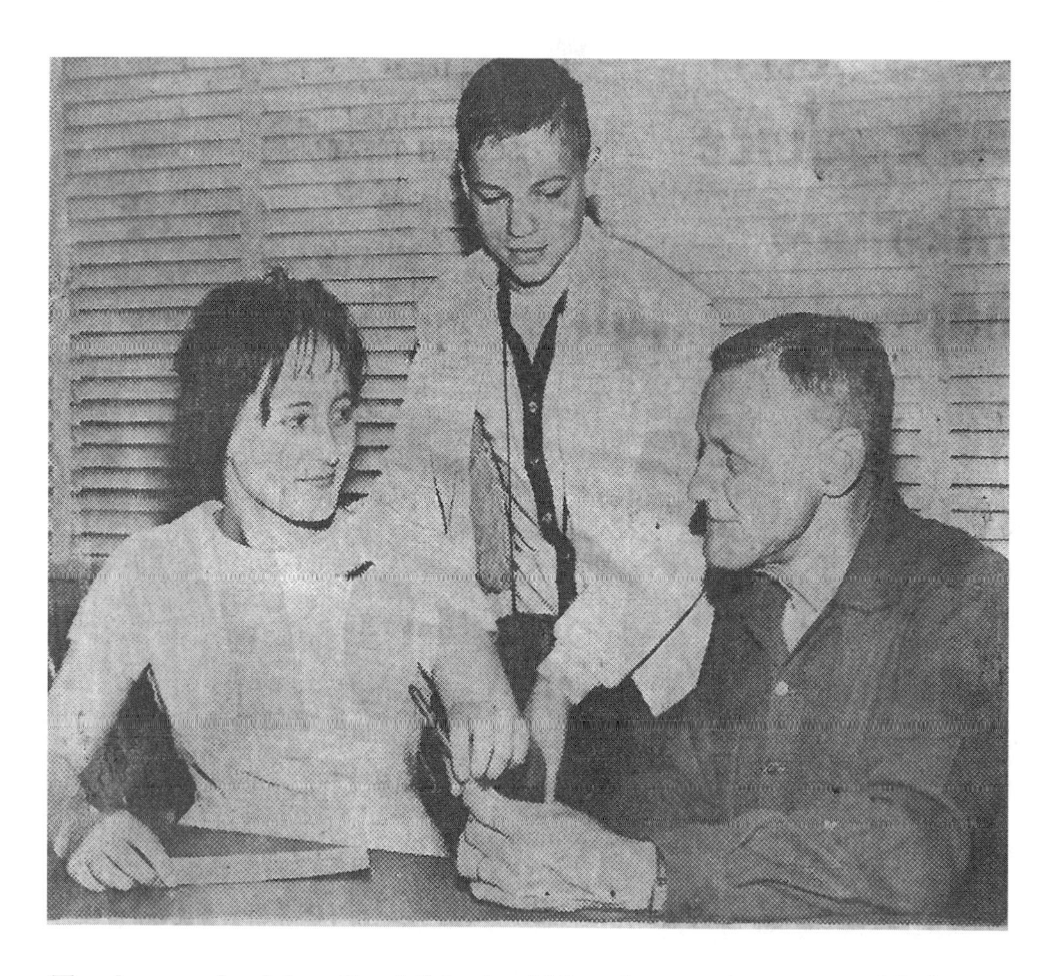

The photography club at Couch School, which no longer exists, presented Jack Marks with a $592 check to keep Belle and the baby at the zoo. Linda Houeland, then fourteen and club treasurer, and Carl Robinson, club president, handed over the check in person. (Reprinted with permission from The Oregonian*)*

a one-eyed cat watching two mouse holes! Not only did I tend and treat animals at the zoo and in my private practice, but now I was a bona fide celebrity. Suddenly I was making speeches at public functions, club meetings, and universities. Heck, I became so popular I practically needed a secretary to juggle my social obligations. Almost daily, I'd wash up after work, climb into a suit, and head off to whatever event, packing my slide projector loaded with all sorts of animal images. I hate to brag, but I was a darn good speaker and often had my audiences in stitches. Mostly, I just enjoyed teaching the public about zoos and animals, and hoped that the knowledge would lead to better animal care everywhere.

On top of everything else, I was an associate professor of animal care at what is now Oregon Health & Science University, making weekly trips there to supervise the medical care of animals. I was also involved in a "think tank" for local school districts, where the question came up: "How can our teachers better educate students about wildlife?" That led to my teaching a continuing education course, "Living Laboratory," three times a week during summers. It was very popular. Teachers had to preregister a year ahead to participate because I limited each class to twenty teachers. We would tour the zoo and talk about each species, everything from nutrition to reproduction. "Matt never passed up the opportunity to teach, to share what he was doing," says my friend Jim Anderson, a naturalist who in the 1960s regularly worked at the zoo. "He enjoyed reaching out to the teachers because he knew they would pass that information along to the children."

Visiting local schools became one of my favorite things to do. The children loved hearing my stories and loved my zoo tours even more. For years I received thank you letters from schoolchildren who so appreciated learning more about animals. I even got letters from children living out of state. In 1967, for instance, a girl named Krista from

I became a sought-after public speaker following Packy's birth. After a long day at the zoo, I'd change into my suit and drive wherever I was invited to share any and all details about Belle, Packy, and the zoo in general. (Courtesy of Dr. Matthew and Patricia Maberry)

Fremont, Nebraska, wrote, "I read the article about the elephants in the Portland Zoo. Is it true that elephants are very afraid of mice? I have heard that they can suffocate an elephant." I wrote back to Krista and said, "Elephants are not afraid of mice any more than any other animal. The problem is that any new smell or sound or sight bothers them until they get used to it, or find out what it is. This is especially true when babies are present. Remember, the only dangerous thing is that which hurts or kills you, whether animal or machine."

I've had many peaks and valleys in my long life, but the months before and after Packy's birth rank among the most exciting in my memory. The experience contributed to veterinary science, and the outcome—a healthy baby—provided Portland with its greatest animal mascot.

I truly wish I could do it all over again.

In the years that followed Packy's birth, I taught a "Living Laboratory" class to Portland elementary school teachers. In the twelve-week course, held after school hours and during summer, I taught teachers about animals so they could conduct their own classroom tours of the zoo. Here, I coaxed Rosy to open her mouth so I could show teachers her four flat teeth. (Reprinted with permission from The Oregonian*)*

Tuy Hoa begged for apples from a first grade teacher from Capitol Hill School during a zoo-site session of "Living Laboratory." The course was part of a new concept in instruction of the sciences in the Portland School District. (Reprinted with permission from The Oregonian*)*

"Dear Doug Baker"

It wasn't unusual for me to get a phone call or a letter about a non-zoo resident needing assistance, like this young girl's seagull. One time I got a call about a seal being lost in the Santiam River. The pet seal, which belonged to a couple traveling through Oregon, somehow slithered out of the backseat of their car into the river at a rest stop. It took a while for authorities to find that seal, but I made sure the couple got their pet back.

Most callers, however, were locals who owned longtime patients of mine. The zoo animals occupied most of my day, but I always made time for the animals I had tended over the years, plus I still accepted new patients as well. Prior to working at the zoo, I had a substantial client-based private practice. I tried to continue to care for those patients when they needed me, which is why I made house calls at night after my day's work at the zoo. Many of these animals, like their owners, became like family to me. If a call came in from a worried owner regarding an ill animal, I got to their home or farm as soon as possible after work or on weekends. It was always a sad day when one of the animals died, both for their owners and for me.

"DEAR DOUG BAKER: In times of strife . . . I thought you might like to know there is a wonderful man among us who performs good deeds and acts of humanitarianism without seeking rewards nor accolade.

"My daughter rescued a seagull . . . drowning at the beach last week. She brought him home to nurse him but found his wing was too severely broken for her to repair. She phoned the zoo to seek advice and Dr. Maberry himself answered the phone and patiently listened to her problem. He explained he . . . could not bring the bird to the zoo because of the rules but asked her whether she was willing to nurse the bird. When she assured him she was, he said he would come to the house after work.

"I was stunned when this important and very busy man came out of his way after a day's work . . . to save one little bird . . . A few days later he even phoned to see how his 'patient' was doing . . . Now there is a big man. Mrs. Jerome Stern, 7300 SW Ridgemont St."

(We have heard similar stories about the zoo veterinarian, who must be a true lover of animals.)

This article appeared in the newspaper after I helped to treat a seagull rescued by a young girl. (Reprinted with permission from The Oregonian*)*

Precocious Packy was off and running from the start. He loved playing pranks on his parents and aunties, and often challenged his keepers—by opening doors and escaping! (Courtesy of Marianne Marks)

Packy & Me, Then & Now

I loved watching Packy grow up, especially his first year of life.

Right from the start, this bouncing 225-pound baby boy proved to be the public's favorite—a very active individual with a great deal of personality. For instance, Packy's complete lack of understanding of what his trunk was for was comical. He would step on it with his foot, and then he would try to pick up things, emulating his auntie and his mother. But his trunk never seemed to go where he wanted it to go!

As the weeks went by he acted more and more like many babies we've all seen — youngsters who play around in their food and so on. We would lay lettuce, cabbage, and other greens on the floor for the adult elephants, and Packy would immediately go over to tromp and roll on them. If there was water on the floor, he would run and jump

The public loved watching Belle and her baby swagger through the elephant yard. (Reprinted with permission from The Oregonian*)*

on it; if there was straw, he'd run, set all feet, and slide. Auntie and Belle seemed to disregard this entirely and did not punish him.

Packy's biggest fun was at night, when his parents and auntie lay down to sleep. He walked along their tails as though they were tightropes; or he'd walk along their trunks, climb on their heads, or kick them in the stomach repeatedly with a hind foot.

Still, no chastisement ever took place—until Packy reached about seven months; then all creation broke loose upon him. When he bothered auntie one day, she knocked him about twenty feet across the floor. A few days later, when he tested and teased his mother too much, she kicked him up against the wall, where he crumbled and fell to his knees. It took him a few minutes to get his breath and get back on his feet, but it wasn't very long before he became a quiet, docile animal who exhibited more respect for others.

Other minor power struggles happened as Packy grew. Sometimes he wanted his mother or the other adult elephants to get up, so he would wander out into the middle of the room and let out a big trumpet. Of course they all jumped to their feet to see what the trouble was. (This trick is so characteristic of human children, too.)

Packy had a tremendous love for his father, and his father had a great deal of patience with him. Even at the age of ten months, this little fellow would go over to play with Thonglaw and try to kick him; his father would simply grab him by the trunk to have a tug of war. But when play became too foolish, his father chastised him rather severely, which Packy seemed to appreciate.

Watching Packy, I learned a great deal about elephants, arriving at the conclusion that they more closely approximate humans than any other animal I have handled. The fact that they are so gentle and so considerate of their young amazes me. It's (continued on page 135)

Playful Packy

This article, written shortly after Packy's birth by reporter Lev Richards of The Oregonian, *provides an insider's peek at how the baby elephant behaved in his first few months of life:*

It's for opening doors.

The other day he turned the knob on the door to the elephant house just like he had seen [elephant] Head Keeper Al Tucker do.

There before him lay the whole wide, wonderful world of kids and cameras and popcorn and peanuts. He stuck his long nose out and started to follow Tucker home.

Then he remembered how mother Belle had paddled him with her powerful trunk the last time he pushed open a door — to the feed room. So Packy just stood there, lollygagging, with one eye on Belle, until one of the zoo patrol caught him in the act and whacked him where babies get whacked.

Now there is an outside latch on the man-type door to the elephant

Packy was a quick learner. It only took a week or two for Al Tucker, his keeper, to train Packy to stand up like a big dog, with his front feet on Al's chest, and his trunk wrapped around his neck. (Reprinted with permission from The Oregonian*)*

Teaching Packy to perform helped keep the mischievous youngster busy. From birth, Packy proved smart, curious, and a little bit sneaky. (Reprinted with permission from The Oregonian*)*

house — out of infant elephant reach.

But no one yet has been able to cure the mischievous elephant child of some of his other tricks — like putting his front feet on the ledge of the window in an effort to rub noses with the children who come week after week to smile and wave and make faces.

Next week Director Jack Marks plans to have horizontal steel baby bars welded to the vertical bars which keep the adult elephant away from the window glass. This will finally make Packy's dormitory Packy-proof.

To compensate for this loss of freedom, Marks has ordered 15 inches of sawdust for the moat around the elephant yard. Belle and her fuzzy-faced baby will then be turned out into the pachyderm patio, early next week.

If the bumptious baby falls into the moat, nothing will be hurt except his feelings.

His hide is thick, but his feelings can be hurt. He sulks if Tucker doesn't give him a love every morning at 7. He's a good baby — sleeps soundly on his side cuddled between his mother's legs. He never cries, except when he's scared or mad. Then he honks like a great gray goose.

He toys with wisps of hay, gnaws idly on carrots, but spits out bananas, apples, and bread when offered. At first he kicked his trunk around with his front feet as if it belonged to someone else.

But elephants are just naturally nosy. Soon little Blue-beard — he still has a ridiculous mess of blue-black hair and a fuzzy moustache — learned to pick up stiff hay with his trunk, and scratch himself.

It is only recently that he has found his mouth. He still smears stuff all over his face trying to find the target.

There was no use putting anything there anyway until now. He is just now cutting his first four baby teeth. They aren't clear through the gums yet.

[In the meantime] he's busy chewing on anything within reach — fingers, shoes, shirts or furniture.

too bad that humans don't have the tremendous love and affection in many instances as the elephant has for its young.

My job at Portland's zoo gave me plenty of chances to observe and enjoy baby elephants. On October 2, 1962, six months after Packy's birth, Rosy gave birth to Me-Tu (pronounced "MEE-too"), so named because her birth received far less public notice than Packy's. Pet delivered Dino ("DEE-no") on September 15, 1963, and nine days later, Tuy Hoa had her baby, Hanako ("HAH-NAH-ko").

Four healthy baby elephants born in eighteen months—amazing since even today one in three births statistically ends in death. Now I had myself a true nursery, as well as a living, breathing tableau for learning. Never before had an American veterinarian been so blessed with eyewitness knowledge of elephant birth.

My knowledge paralleled strides in veterinary science, which really took off in the 1960s, but not as fast as it would today. Modern conveniences like computers, fax machines, and cell phones either didn't exist or were in their infancy. So it took awhile for zoo veterinarians to share what they learned. It wasn't until 1983 that the American College of Zoological Medicine formed to certify veterinarians with special expertise in zoological medicine.

The public also became better informed about exotic animals, in large part due to television and motion pictures. Suddenly animals only seen in *National Geographic Magazine* lumbered or flitted through family living rooms via shows like *Zoo Parade* in the 1950s, and *Wild Kingdom* in the 1960s. The 1966 blockbuster movie *Born Free*, about a husband-and-wife team who raised a lion cub to adulthood and returned it to the wild, emotionally enlightened the world about wildlife.

I guess I was on the cutting edge of change, because the deluge of queries after Packy's birth came from veterinarians and physicians everywhere, all wanting to know more. Some visited the zoo, while

Packy's Son Paints

Packy's son Rama has been taught to paint as part of an ongoing enrichment program that provides elephants activities and mental stimulation at today's Oregon Zoo. One of Rama's vivid paintings—described by an art critic as "abstract eruptionism"—decorates a hallway in our home. Those who purchase a Rama painting also receive a certificate of authenticity that is stamped with a "trunk print" of Oregon's "biggest" artist. To learn more about Rama's art or to purchase a painting, go to www.oregonzoo.org/Rama/index.html.

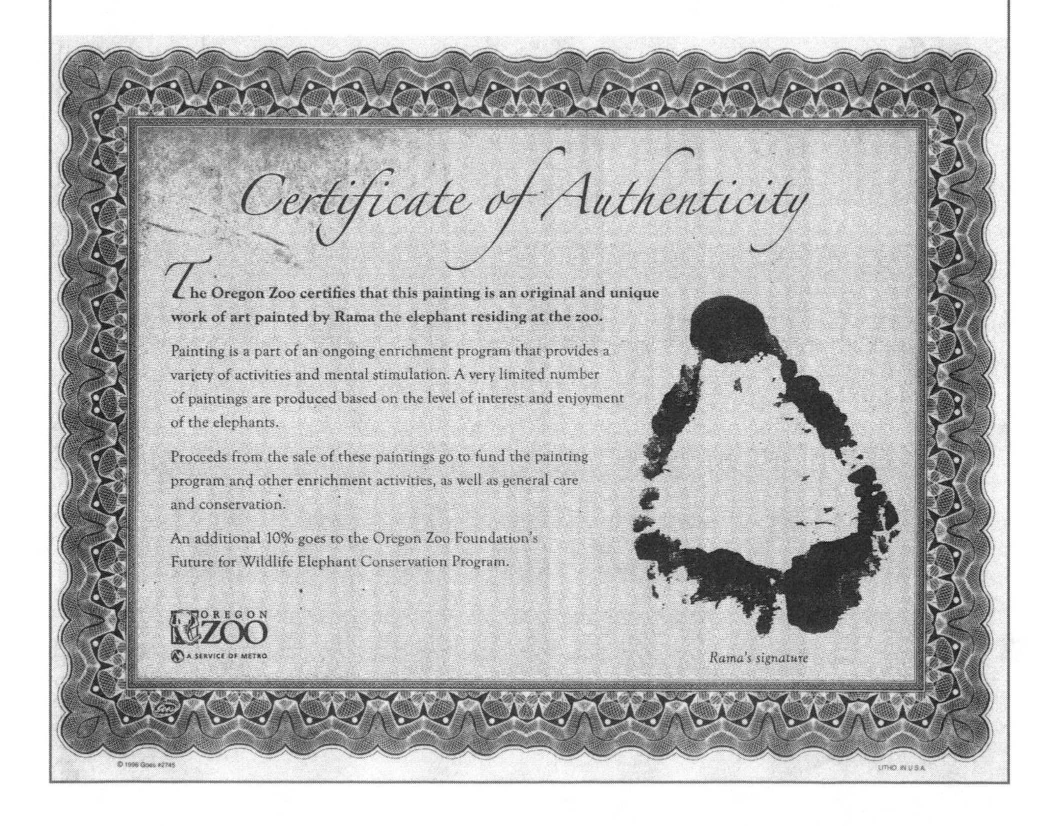

Certificate of Authenticity

The Oregon Zoo certifies that this painting is an original and unique work of art painted by Rama the elephant residing at the zoo.

Painting is a part of an ongoing enrichment program that provides a variety of activities and mental stimulation. A very limited number of paintings are produced based on the level of interest and enjoyment of the elephants.

Proceeds from the sale of these paintings go to fund the painting program and other enrichment activities, as well as general care and conservation.

An additional 10% goes to the Oregon Zoo Foundation's Future for Wildlife Elephant Conservation Program.

OREGON ZOO
A SERVICE OF METRO

Rama's signature

© 1996 Goes #2745

LITHO IN U.S.A.

others actually went there to conduct research. Now that I had deciphered the secret to elephant birth, the veterinary and medical communities scrambled to build on that knowledge.

Of course I supported further research on elephants—heck, all animals! Veterinarians knew damned little about most wildlife, elephants included. Better knowledge, I believed, would lead to better care. Yet, I wasn't sure the zoo was necessarily the venue to do that research. The zoo is built and supported by taxpayers who want a place where they can enjoy and get to know animals. Tax money, I felt, should not be used to fund a private laboratory for professionals with special interests—looking, perhaps, for the same publicity Packy's birth had generated.

Well, the more scientists became interested in what was going on at the zoo, the more political the zoo got. Things really got sticky in 1971, when Packy turned nine and the Portland Zoological Society took over the zoo's management from the city of Portland. Jack Marks, who had fought hard for the construction of the new zoo, retired and an outsider named Philip Ogilvie replaced him. Many zoo employees subsequently left, concerned—maybe confused—by the zoo's new direction. Others were laid off or fired.

Mike Keele was a young employee at the zoo during this rocky period in its history. Twice during Ogilvie's four-year tenure, he says the zoo went bankrupt.

I can't say that I got sick of things and quit (though Ogilvie's management greatly disturbed me). Rather I got sick and had to stay in bed for a year. In 1973 an African green monkey was allowed illegally into this country and ended up at the zoo, where it infected me with a highly contagious, shingles-like herpes virus. All sixty animals in the zoo hospital at the time had to be euthanized. Tests showed that I actually was infected with five types of herpes, which I had contracted

PORTLAND, OREGON | PORTLAND ZOOLOGICAL SOCIETY | FEBRUARY 1968

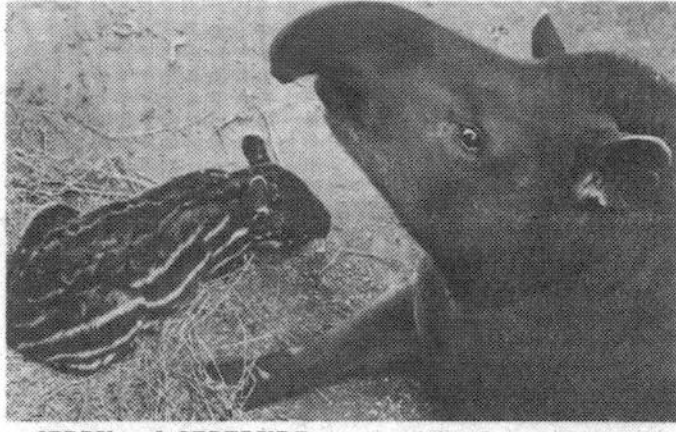

JERRY and GERTRUDE, a pair of South American tapirs (tapirs are the largest South American mammal) purchased by the Zoo Society as a gift to the Zoo, gave birth Feb. 1 to their second infant. Striped at birth, the young tapir will lose his markings as he matures. Tapirs are pachyderms, sometimes described as "pigs that started out to become elephants and changed their minds." They have a short trunk and 14 toes — four on each front foot, three on each hind foot.

AT LEFT: Dr. Matt Maberry, zoo veterinarian; Lieut. Willey's wife, with photo of her husband, and Zookeeper Larry Koppe get good look at new python.

Python from Battle Zone Changes Skin AND Scene

Lieut. William A. Willey's second platoon took a strange prisoner while on patrol north of Long Thanh in December — a reticulated python, 16-plus feet long, weighing close to 100 pounds.

Willey, a Portlander, wrote Zoo Director Jack Marks:

"We packed him 7 miles, wrapped in three ponchos, to our next stop.

". . . We would like to keep him," he added, but combat conditions made it impossible; "if we turn him loose, the Vietnamese in the area will kill him for dinner."

After getting assurance from Marks that "The Big Worm" would be welcomed at the Zoo, it was taken to Saigon by helicopter, then to Portland by jet. It's shedding its skin, too.

Reticulated python, largest of the constrictor snakes, may attain a length of 30 feet. They range in Burma, Indo-China, and the Malay Peninsula and archipelago. They kill their prey by constriction — by wrapping it in the coils. Death is by suffocation; bones of victims are seldom broken; prey may even be swallowed alive.

Promptly named "Willey Jr.," the Viet snake was placed in an exhibit with another Zoo python, Bettina. Willey caused a minor flurry of excitement by turning out to be a "she," but it didn't last long. Bettina is a "he."

EXPEDITIONS *with Jim Anderson*

EXPEDITIONS into the Oregon outdoors begin again with Spring; call 226-6501 or 226-7639 to register. Coming up:

MARCH 1-3: 8th graders and up, spend weekend at Warm Springs Reservation with conservation officer at Agency; hear about and see range timber and wildlife conservation work; seek Golden Eagles in Lake Simtustus and Lake Chinook areas. Leave 6 p.m. Friday, return 9 p.m. Sunday. Cost: $8, members; $10, non-members of Zoo Society.

Jim Anderson

MARCH 9: 5th-7th graders go on "Shark Teeth Hunt" at site of ancient ocean bay, dig remains of primitive fish. Leader, Mike Goodrich. Depart via Zoomobile from Children's Zoo at 8 a.m., return 6 p.m. Cost: $4, members; $5, non-members.

APRIL 6: 5th-7th; "Spring Opener" to see fresh water animals in western Multnomah and Washington counties.

APRIL 13: "River Gate Wildlife Trip," 5th-7th graders, hike through area at confluence of Columbia and Willamette to observe waterfowl, small mammal and owl population.

APRIL 20: 8th and up, "Major Creek Herpetelogical Field Study;" *APRIL 27:* 5th-7th, Garter Snake Hunt; *MAY 18-19:* Owl banding trip in Central Oregon.

It wasn't uncommon to see me in Zoo News, *which was published by the Portland Zoological Society. Here, I'm holding the tail of a sixteen-foot-long python captured by an American platoon in Vietnam. Lt. William A. Willey, a Portlander, wrote Jack Marks and reported, "We packed him seven miles, wrapped in three ponchos, to our next stop." After Jack agreed to take "The Big Worm," the python was flown by helicopter to Saigon and then by jet to Portland. We named the python "Willey Jr.," then learned he was a she. Oh, well . . . (Courtesy of the Oregon Zoo)* from treating animals over the years. But it was a baboon herpes, transmitted by the green monkey, that nearly killed me.

Most people do die from this virus, which is extremely painful. And I had suffered severe pain before in my life; in my teens a tractor accident maimed my back so badly the doctor told me I would never walk again. He was wrong, but that injury kept me out of World War II and plagued me throughout my life. Nothing, though, compared to this pain and its viselike grip around my rib cage. I thought—no, I prayed—I would die. Gradually, however, I recovered, largely because my Japanese doctor gave me small doses of smallpox vaccine.

That's when I got a visit from Ogilvie. It struck me as sort of funny, him coming to visit, since the Zoological Society originally offered me Jack's job. But I'm not a desk job kind of guy; I need to be outdoors, free, on the move. So Ogilvie got the director's job, which meant he was my boss. And in 1973, while I lay in my sickbed recovering from a zoo-inflicted disease, Ogilvie and the Zoological Society president visited me at my home and asked me to retire. After consulting my physician and considering my poor health at the time, I decided to pursue other opportunities. My veterinarian friend, Dr. Richard Werner, was interested in my previous position, and would have been a fine zoo vet, but it didn't work out with the zoo. Fortunately, we have remained close friends all these years.

Looking back, I must say my health worried me more than my job. But I couldn't help feeling sad and somewhat angry. I loved my animals; I still miss them.

After I recovered, I just wanted to get away from Portland and begin a new life, so I moved to Southern California and took a job working for the federal government in animal welfare. Basically, I inspected animal research facilities between the Mexican border and Bakersfield, and shut down those violating regulations.

In 1974, though, I moved back to Portland; my dear mother was gravely ill. It's a helpless feeling, your mother dying. She was eighty-six and had lived a long, productive life, but I couldn't help feeling I had not done enough. She passed away later that year.

Fortunately—happily—my move back to Portland ended well. In 1975 I married my best friend, Patricia, a slender spitfire I had met in 1971 and asked to volunteer at the zoo. Until then (she'll tell you), I had a way with the ladies. They flirted, I flirted. Occasionally, busy schedule allowing, I dated. But no woman could keep up with my crazy life.

Until Patricia.

Twenty-five years my junior, she not only kept up, she enjoyed the chaos. By day she was a well-dressed buyer at Meier & Frank. At night she was my gritty, able assistant, willing to leap out of bed at 2 AM to tend whatever animal needed help. After marrying, Patricia, like me, slept with her clothes by the bedside, always at the ready.

Now I've done all sorts of things to heal animals, none that I particularly want to brag about. But Patricia, if asked, will tell the story about the white cat that clarified her feelings for me. "While I worked with Matthew at the zoo hospital, someone dropped off a stray white cat with a badly broken leg," she'll begin. "This cat fascinated me since it had one blue eye and one brown eye. Matthew, though, was very puzzled about the best way to set the leg, considering that he wanted the cat to be mobile as soon as possible. Finally he asked me to find a coat hanger. Then, as I held the cat, Matthew took a pair of pliers and twisted and turned and shaped the coat hanger to fit the cat's leg. Finished, he wrapped the leg in that coat hanger with tape, and then told me to put the cat down on the floor. After one trepid step that cat was off running through the clinic.

"I don't know what happened to the cat, but I do know what

happened to me. After watching this famous and skilled veterinarian spend so much time and worry on this little stray creature, I became aware for the very first time how deep my feelings were for this man."

I didn't propose in a very romantic way. After moving back to Oregon, I had reconnected with an old friend, Bob O'Loughlin, who exhibited exotic animals in trade shows. We formed a company, M.O.L. Enterprises, which combined our names, Maberry and O'Loughlin. It was my job to research and find possible animals for import, and Bob exhibited them. A few days before one animal research trip to Asia, I told Patricia I was going and that I'd like her to come along. And if she wanted to go as my wife, that would be just fine.

I have no idea why, but Patricia said yes. We married in May 1975, when I was fifty-seven and she was thirty-two, and nine people attended our small wedding.

"I didn't realize how famous [Matthew] was until I went to get the marriage license," Pat said in a 2008 article in *The Oregonian*. "The clerk asked me: 'Is this the Dr. Maberry who delivered Packy?' When I said yes, she told me we didn't want the announcement in the newspapers. She knew the public would be all over it."

Once wed, Patricia and I became a team and formed our own business, Wildlife Enterprises, specializing in the care of domestic and wild animals. Daily I drove 150 miles or more in my Chrysler, packing my heavy leather bag and a veritable traveling pillbox in my trunk, doing surgery and treating animals. Many clients felt like family, I'd known them so long.

We also kept busy touring the world for M.O.L. Not all animals I found came back to Bob for exhibit. One project for the National Institutes of Health required that we obtain rhesus monkeys from the primate-packed jungles of Bangladesh. Scientists took a kidney from

The seals are the main attraction at the Seaside Aquarium, where I've been the official vet for about thirty years. Previously a natatorium, the aquarium is supplied by water piped in directly from the ocean, which allows it to create a more natural environment for its critters. Around since the 1930s, it remains a neat little mom-and-pop treasure, and the seals love to perform for their fish treats from visitors. (Photo by Brandy Hussa, courtesy of the Seaside Aquarium)

each monkey for the production of the polio vaccine. Every monkey lived, minus a kidney.

In 1978, though, I suffered a heart attack, requiring bypass surgery. It slowed me down for a time, but once I healed, I kept busy consulting and helping clients prepare animals for shipment back to their natural habitats. I continued to speak at public and private functions as well, and I appeared on television as a guest veterinarian, discussing the care of pets.

Also, every month for the last thirty or so years, Patricia and I have motored to the Seaside Aquarium where, as the "official," government-appointed veterinarian, I check the federally protected seals for health issues and conduct postmortems at death. General Manager Keith Chandler has overseen the aquarium for many years. Both he and Jason Hussa have assisted me during my monthly inspections of the seals, and also with any needed medical procedures. "It is Doc's compassion that sets him apart, not only . . . to the staff

General Manager Keith Chandler and I have worked together many years at the Seaside Aquarium, where I check on the seals monthly. (Photo by Patricia Maberry)

of the aquarium but, morc importantly, to our beloved harbor seals," says Jason. "Even though he lives nearly two hours away, he's very much the 'vet next door.'"

Of course, I often think of my old friend Packy. Like Thonglaw in the 1960s, Packy's now the patriarch of the Oregon Zoo's elephant herd. Nearing age fifty, Packy stands a hulking ten feet, six inches tall, and weighs an astounding 14,500 pounds, making him the oldest and largest Asian elephant in North America. He's also the only second-generation captive elephant to father children; he's had seven, though several died shortly after birth.

Packy's birthday, April 14, is still a huge (continued on page 147)

In 2007, during the filming of a segment for Oregon Field Guide, *our friend Sue Anderson captured this fitting portrait of Packy and me at the Oregon Zoo. (Photo by Sue Anderson)*

Every year since Packy's birth in 1962, a birthday party has been held at the Oregon Zoo. Here the zoo's kitchen staff prepares Packy's cake, complete with raw carrots jutting from the frosting. (Reprinted with permission from The Oregonian*)*

Packy always demolishes his birthday cake, to the delight of thousands of spectators who annually pack the zoo for the special celebration. (Reprinted with permission from The Oregonian*)*

Before her death in 2005, writer Shana Alexander—who was the lone female reporter during the Great Portland Elephant Watch—kept in touch with letters on a regular basis, and visited Patricia and me in our home several times. I always considered her a remarkable woman. (Photo by Patricia Maberry)

Portland event. Thousands pack the zoo annually to watch him eat his carrot-studded cake and raise his trunk in greeting. Always, the crowd roars back its love.

After I left the zoo in 1974 I never went back to visit; it didn't feel right. But in 2002, when zoo officials asked me to attend Packy's fortieth birthday, I was finally ready. It was a pleasure to attend the event and see all the positive changes at the zoo. It especially touched my heart when a beautiful seven-year-old girl came up to me and personally thanked me for delivering Packy.

Jim Anderson (left) was with me the day that cameras rolled in the elephant barn for Oregon Field Guide. *(Photo by Sue Anderson)*

Since then, once or twice a year, Patricia and I visit Packy and his current pals. Unfortunately, Belle, Thonglaw, Pet, and Tuy Hoa have passed away, but the current elephants—Packy; his son, Rama (pronounced "RAH-mah"); his daughter, Shine; Chendra ("SHEN-drah"); Rose-Tu ("Rose-TOO"), and her two-year-old, Samudra ("Sah-MOO-drah")—always welcome us like old friends.

Samudra's birth, in fact, resurrected my 1962 self. Rose-Tu, daughter of Me-Tu (born to Rosy six months after Packy's birth), was quite pregnant and scheduled to deliver in August or September of 2008. But an elephant had not been born at the Oregon Zoo since 1994; the zoo, like zoos nationwide, was full of captive-born elephants, so

In my nineties, I'm still very much on the go. Patricia and I are involved in many organizations, and nearly every afternoon, I head out the door for coffee with friends. I also enjoy restful moments with our poodle, Zandi, who loves to nap with me in my recliner. (Reprinted with permission from The Oregonian*)*

breeding mostly stopped. Most of the Oregon Zoo's current elephant keepers had never witnessed or assisted at an elephant birth.

Our good friend Mike Keele asked me to come to the zoo to share with staff what I knew about elephant birth. So in May 2008, I sat in the elephant barn, and "held court," as Patricia likes to say, explaining to staff and volunteers the details of Packy's birth. As it turned out, baby Sam's birth in August 2008 was a little rocky; Rose-Tu, like some elephant mothers who have never witnessed a birth, harshly kicked her newborn and refused to nurse. Within a day, however, Rose-Tu

Packy is now the patriarch of the Oregon Zoo elephants. (Photo by Sue Anderson)

knew her role, and much to the relief of the zoo staff, mother and son successfully bonded.

Just like Belle and Packy did.

The year 2008 was a big one for the zoo. First came Sam, who generated almost as much public adoration as Packy. Then voters approved a $125 million bond measure that, among other improvements, will boost elephant living space from the current 1.5 acres up to 6 acres. I applaud the creation of this elephant reserve and told voters so at a preelection meeting of the Metro Council, the Portland area's tri-county agency that now oversees the Oregon Zoo, among other public entities.

Also in 2008 I received three lifetime achievement honors: the first from the Oregon Veterinary Medical Association; the second from Metro Council; and the third from my alma mater, Washington State University. It felt pretty good to get such recognition.

These days, in my nineties, my body's slowed down but my mind still whirs along. Patricia, thankfully, keeps me busy attending meetings or visiting friends. Sometimes, though, I just relax in my recliner, often with our beloved black poodle, Zandria, on my lap. She's 16½ years old—that's 115 in human years!

Well, that pretty much wraps things up. Hopefully you have learned a thing or two about elephants. Goodness knows I've learned enough to last a lifetime. And I've been fortunate to have a long life, one that many might envy. I've done things that most men only dream of.

And you know what they say—that elephants never forget? I'm sorry I can't verify that fact. One thing's true for me, though.

I'll never forget Packy.

Good-bye, Dear Friend

Morgan Berry died in 1979 under bizarre circumstances. Some theorize that it might have been a heart attack; the week before his death, Patricia and I visited him and he couldn't walk ten feet, he was in such ill health. Then again, maybe it was just a tragic accident. All we know for sure is that when Morgan's son Ken found his sixty-eight-year-old father, his body was barely recognizable.

Writer Shana Alexander visited Elephant Mountain shortly after Morgan's death and interviewed neighbors Mary and Joe Wodaege, whom Morgan called nightly, to let them know he was okay. When he didn't call one night, and could not be found during a late-night search, the Wodaeges called Ken, who lived in Seattle at the time.

"Next morning," Joe told Shana, "Kenny come by our place around six or seven o'clock, and when we all got up to the farm, I noticed Buddha [the elephant] was lying down. We checked the barns first, didn't find nothin', and when we come out Buddha was up on his feet, kinda workin' toward us, occasionally lunging, and his trunk sorta lashing and coiling up on itself like a python. I'd never seen anything like it.

"Then we seen this object that we hadn't noticed before lying on the ground. It looked like an old rag—no, maybe more like a deer hide. Kenny got a pitchfork and raked this hide out and started to unfold it. He unfolded the first flap and two strings came out and they was legs. Then two more strings come out and they was arms. Unfolded it a third time and we saw this face. It was Morgan's profile. A perfect image but absolutely flat, no eyeball even, and that's the first time we even knew what the thing was. Kenny said, 'That's my father.' Then he said, 'Buddha, how could you do this!'"

Some people believed that Buddha stomped Morgan to death, but not me. Elephants rarely trample anything. They're more likely to tear something up, or lean on it, or hit it with their trunk. They'll even do a headstand on you. But they're very fussy about where they place their feet. As I told Shana, "You gotta remember that an elephant just leaning on you in love leaves nothing but a grease spot. A man is a very fragile creature."

Ken called the authorities, but it took awhile to recover the remnants of Morgan's body; there wasn't enough of him left to conduct an autopsy. Protective Buddha, who may have been in musth the night of Morgan's death, charged anyone who came near. Finally Ken told officers to shoot and kill Buddha.

After the authorities recovered Morgan's body, they had other worries. Thai—the same elephant who disappeared after Eloise Berchtold's death—was missing. Cowlitz County deputies soon found Thai on a nearby road. Officers blasted their siren and flashed their lights, hoping to force the elephant to retreat, but three-ton Thai charged the annoying car instead. The terrified officers rammed their patrol car into reverse, flipped around, and sped away to safety.

Eventually, Portland zoo officials showed up and shot Thai with a body-numbing tranquilizer dart. Groggy but still able to walk, Thai allowed himself to be led back to Morgan's farm.

That day ranks among the darkest days of my life. I knew a lot about animals, but Morgan knew everything about ele-

Patricia and I loved visiting Morgan and his family on Elephant Mountain. It was like Africa—in southern Washington. Animals of every size wandered freely, as nature intended. (Courtesy of Dr. Matthew and Patricia Maberry)

phants. He was a brilliant man who helped unravel the riddle of captive elephant breeding. Packy and hundreds of other elephants worldwide might not be here today if it wasn't for Morgan Berry.

You know, the last time I saw Morgan, he must have known his time was near, because he told me, "Someday they're going to find me out there with those elephants. And that's how I want it."

I'm glad Morgan got his wish, but I wish more that he was still around. I miss my good friend.

Afterword:
A Life That Most Men Only Dream Of

I've had many adventures and learned many new things in my ninety-plus years of living; so much to do, so little time.

Elephants dominated my life for many years, and Packy's birth just made me more curious about the breed and how they reproduce. But other interests snagged my easily bored mind. In another life, I might have been a noted boatbuilder, a gutsy pilot, or a medal-winning baker.

In this vet's life, I am a mere hobbyist.

One skill I excel at, though, is storytelling. I'm a reader and a thinker, which I suppose makes me a natural at spinning words, verbal or written. And I've always noticed how everybody loves a good story.

Well, here are a few more of my stories, most beyond my life with Packy. They sure made good memories for me. Hope you enjoy them, too.

The Finnish Family

I begin at the beginning, back in the late 1940s, at my first job as a professional veterinarian in Everett, Washington. One day my new boss said, "Matt, I want you to go out and visit this Finnish family. Go and clean their cow. They've kicked every veterinarian off their property. Tell them I'm too busy to get out there."

What could I say?

I visited the Finnish family.

I was met at the gate by a huge Finnish woman who curtly asked my identity and then, arms crossed, declared: "I don't want you."

Well, lady, I thought, I don't want you either. "Cleaning a cow" is dirty business; I had to stick my whole arm into a deep hole I preferred to avoid in order to pull out afterbirth that hadn't come out naturally. If it was not removed, the cow could die.

Truthfully, I replied, "I'd rather be in the office than doing this dirty job."

Then the Finnish woman, arching an eyebrow, threw me off guard by asking, "What nationality are you?"

I was sharp enough to realize that if I said Finnish, the barnyard gate would open. But my British Isles blood stopped me. So did all the other smaller European bloodlines coursing through my body.

"I'm pretty much a mess," I said.

Well that charmed her. She complimented my sense of humor, opened the gate, and barked, "Go up and see the old man at the barn, and see if he will let you take care of the cow."

When I got to the barn, I saw the old man sitting on a stump, whittling. He never looked up. So I picked up a piece of wood, sat on another stump, pulled out my knife, and started to whittle, too. Finally the old man said, "I see you got by the old lady. I suppose you want that cow in the barn."

"I don't aim to foller her around the field," I replied.

The old man got the cow, then sat on a stool and told me: "I can do this," meaning clean the cow. That set me off.

"Why in the hell didn't you?" I asked. "This is one of the most risky jobs that I could ever be doing. I have friends who are paralyzed from cleaning cows."

The old man just shrugged, so I got to work.

When I was done, the old man said, "I suppose you're so busy you don't have time to go behind the barn."

"Sure I have time," I said.

Behind the barn, I discovered an absolutely beautiful garden. Better yet, standing there grinning was the old man, his arms bursting with fresh-picked vegetables—*for me!*

"The old lady has your check on your way out," he said.

At the gate, the huge Finnish woman stood there, gripping the promised check—and the largest bouquet of flowers I have ever seen.

Later, when I walked into the office, my boss said, "I don't want those flowers." And I said, "They're not for you, they're for me." Then I handed the check to my boss, who, quite stunned, said: "I'll be damned. They never treated me that way." I just smiled.

After that, once a week, the Finnish family called and paid me to come and see them. Most times, their animals were plenty healthy. The couple just wanted to visit—sit, talk, and drink cup after cup of coffee.

Eventually, career-wise, it came time for me to move on from Everett. I went to tell the Finnish family, and they both started to cry. Then the old man gave me a knife that his Finnish brother had made—to remind me of our friendship and of his appreciation.

I still cherish that knife.

Boatbuilder

In the early 1950s I built a boat.

I'd been up to my brother's place, and he had already nailed together a partial framework. I guess the project annoyed him, because when I asked him what he planned to do with it, he said, "Burn it."

That cinched it for me. "I'll take it and I'll finish it," I said.

I didn't know a thing about boats, but I liked the idea of building one. So I brought that boat frame down to Portland and worked on it for probably a year and a half. It was a twenty-one-foot cabin cruiser, built of Port Orford cedar, which you can't get anymore. I may have been a novice builder, but it turned out to be a good boat, and several times I took it out on the ocean.

After awhile, though, I was too busy to own a boat and gave it away to a friend, who was overjoyed. The last time I saw that boat was in 1960, two years before Packy was born, but I've since thought about it many times. I hope it has cruised endless waters.

Pilot

As I mentioned earlier in the book, I tried to get into the Army Air Corps during World War II, but they turned me down because my back was killing me. One day when I was sixteen, I was using a tractor to pull out a tree stump when the stump upended and headed straight at me. The tractor, fortunately, got the brunt of the impact, but my body got the rest. I ended up with a herniated disc. The doctor said I would never walk again. But I did.

I didn't give up on flying either. Before the excitement of Packy's conception and birth, I took flying lessons and got my license. I even bought an airplane—a Mooney four-seater. I flew that plane mostly

Back in the 1960s I often flew my Mooney four-seater around the state and beyond. (Courtesy of Dr. Matthew and Patricia Maberry)

out of Aurora State Airport (thirty minutes south of Portland), where it cost only $5 a month for a tie-down spot. I flew to Canada, to eastern Oregon, to Eugene.

I just loved being up there, soaring like a bird.

Whales

Sea life has always fascinated me, and I especially love whales.

In the 1960s, while I still worked at the zoo, I joined several seafaring teams in search of killer and beluga whales. I went on expeditions with my good friend Bob O'Loughlin to Naknek, Alaska; (continued on page 163)

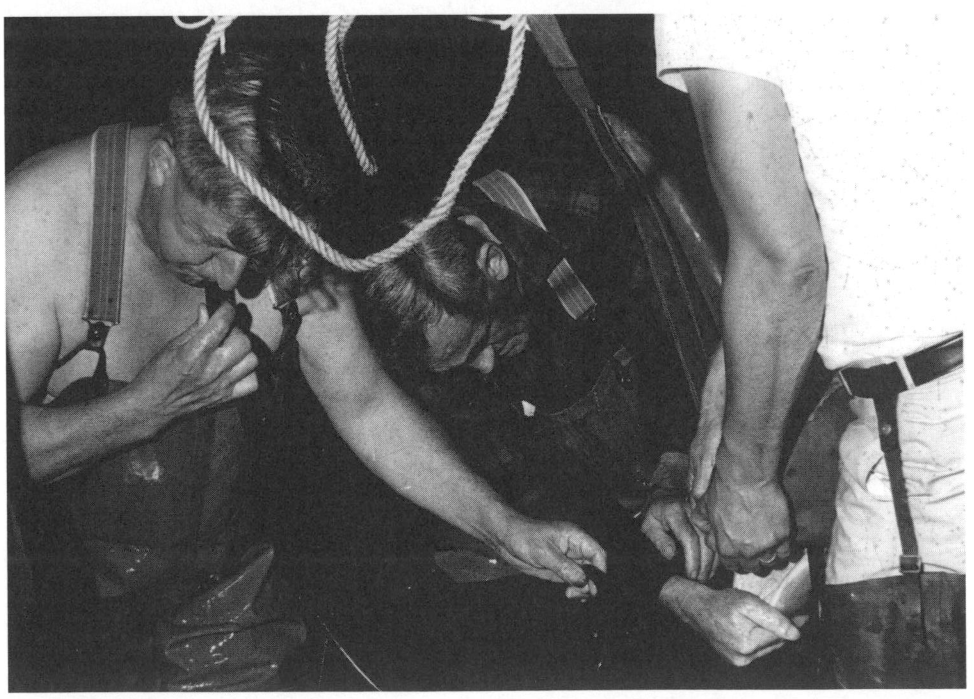

(Above) I went on many whale expeditions, hoping to find a killer whale for exhibition. On one trip were, from left, Tom O'Loughlin, Bob O'Loughlin, me, and Lee Estes. (Courtesy of Dr. Matthew and Patricia Maberry)

(Opposite top) John O'Brien, center, co-owner of the Seaside Aquarium with Bob O'Loughlin, helped me to acclimate a beluga whale (planned for exhibition) in the aquarium's whale tank. The animal, one of three whales who experienced complications in transit, ended up being the lone survivor. (Photo by Jim Anderson, courtesy of the Seaside Aquarium)

(Opposite bottom) Until the whale was stabilized, I had to force food into him with a feeding tube. (Photo by Jim Anderson, courtesy of the Seaside Aquarium)

2 J 3M THE OREGON JOURNAL, MONDAY, FEBRUARY 8, 1965

Wanted: Killer Whale - - Alive

By TOM McALLISTER
Journal Outdoor Editor

Wanted!

A Pacific killer whale about 14-feet in length and weighing 2,000 pounds.

And the critter has to be taken alive.

Out to fill this order — and they've been quietly working on it for the past three months — are four Portlanders, each with his own reason for acquiring a killer whale.

Brothers Bob and Tom O'Loughlin want to exhibit one of the strikingly marked killer shales when they put on the Portland Boat and Sport Show Feb. 19-28.

AFTER THAT, a healthy, captive, killer whale would be a prize dickered for by marine aquariums that have been trying to capture one. These include Marine Land of the Pacific.

Dr. Matthew Maberry, veterinarian at the Portland Zoo, is fascinated with whales and is providing technical knowledge that could make possible the capture.

Tranquilizer charges to be shot into the whale have been devised by Maberry, with assistance from the University of Oregon Medical School.

Gunner for the whale - capture enterprise is Lee Estes, a custom gunsmith who has developed two shoulder weapons that differ from any ever used in shooting tranquilizers or in harpooning seals or whales.

ESTES won't say anything, for now, about specifications of the guns. There is other competition in the pursuit of killer whales, and this team is out to win first.

Only one killer whale has ever been held captive, and this by chance when one was harpooned off Vancouver Island last summer in order to obtain a dead museum specimen. In spite of the wound, the whale was frisky, so it was played out and towed to Vancouver, B.C. for display. "Moby Doll," as it was called, lived a few months.

The Portland team would mark their killer whale. One of the desired size would be picked from a pod. They travel in family groups with the larger herd bulls riding to the outside in protective fashion.

IF THEY knocked out a big killer whale it would do the team no good, because they could not handle anything reaching the 25-foot length and 8-ton weight of the large animals.

Estes is aiming for a 2-year-old whale that is weaned from its mother.

When the right whale is picked, Estes — hovering over the pod of whales in a helicopter — has three seconds in which to hit the whale with a special marker harpoon so it can be tracked through the ensuing chase.

Maberry points out that one full transquilizer shot can't be given the first time Estes fires. If a full one were administered the whale, when it sounded, would pass out under water and perish. The trick is to put the whale to sleep slowly in a series of shots.

ESTES will shoot the marker harpoon first and aim to hit alongside the tall dorsal fin. The harpoon is made to bury only superficially into the fat and muscle of the back. To a light nylon line which it carries, a halibut buoy is attached.

QUARTET from Portland now whale-hunting in Canadian waters includes Dr. Mathew Maberry, top left, Tom O'Loughlin Jr., Bob O'Loughlin, bottom left, and Lee Estes.

The only thing Estes will say about his tranquilizer gun is that he found the standard gas-propelled guns too unreliable and short-ranged in carrying the tranquilizer dart.

Now, with his own handloaded shells propelling the dart, he gets uniformity and can put the dart consistently into a 4-inch circle at 50 yards.

ASSUMING they can knockout a killer whale, the team would then cradle it over an inflatable rubber boat, put halibut buoys under the flukes and tow it to port, where it would be loaded aboard either a truck or onto a boat for shipment to Portland.

The whale must be closely attended and water must be poured over its body, its temperature held even and the crate must be tilted from side to side every 15 to 20 minutes to help its breathing. Otherwise, the weight of the inert body would collapse the lungs.

The media covered at least one of the whale hunts I went on with Bob O'Loughlin and the gang. (Reprinted with permission from The Oregonian*)* the Bering Sea; the North Coast of British Columbia; and the San Juan Islands. Exciting and dangerous, it still thrills me to think about how I bobbed in the ocean eye to eye with a whale, the most intelligent animal I've ever seen.

Mostly on these expeditions I provided medical knowledge and animal care skills, but occasionally I also got to pilot the boat and follow the whale pods. (I also established the first-ever blood parameters on the white whale in the Bering Sea, knowing such information would help veterinarians treat whales.) Finally, in 1967, when I was on an expedition with private businessman and adventurer Ted Griffith, we netted one of the first killer whales in captivity. That whale, eventually named Namu, ended up at SeaWorld in San Diego.

On another adventure, I helped capture four young beluga whales off the Aleutian Islands. During the return airplane trip, one whale went into cardiac arrest, and I had to think quickly to save him.

"The good doctor became legendary at 10,000 feet," remembers my buddy Bob O'Loughlin, who was on that trip. "It was a sight to see, watching him give mouth-to-blowhole resuscitation to that whale."

The whale, incidentally, lived.

Disney

Back in the late 1960s, I was notified that the folks at Walt Disney Productions were attempting to take photographs of a silver fox in Santa Cruz, California, but the silver fox wasn't showing up well on film. They asked me to come to Santa Cruz and make that fox look more silver. I went and gave the fox a bath in alcohol and ether, which dyed its fur silver. I have to say, that fur came up beautiful in the photographs!

Another time, the Disney people wanted to photograph a raccoon, so I took this raccoon to Northern California. I turned the raccoon loose

Outreach and education are a big part of Jim Anderson's mission, as you can see from this 1969 presentation he did with an owl at OMSI (the Oregon Museum of Science and Industry). Jim, a naturalist, actually worked full-time at OMSI but helped out a lot at the zoo, especially at the Children's Zoo. (Courtesy of Jim Anderson)

into some big timber so they could take pictures of it in the big logs. I have no idea why they wanted those pictures, but it was a fun day.

The Owl Is in the Batter

On the rare day when I had spare time, I often headed to the Portland home of my good friend, Jim Anderson. Jim and I frequently worked together at the zoo and especially in the Children's Zoo, which opened

Jim Anderson and I have been on many adventures together since our early days at the zoo. This image is from our day of filming at the zoo for Oregon Field Guide. *(Photo by Sue Anderson)*

inside the regular zoo the year after Packy's birth. Jim also worked at the Oregon Museum of Science and Industry, located next door to the zoo until 1992. (Now the museum sits on the east side of the Willamette River just south of downtown Portland.)

Jim and I both liked to fly, in his plane or mine, typically to rescue sick or injured animals and birds. I would doctor the animal and maybe take it back to the zoo for treatment, or Jim would bring the patient home for rehabilitation. His book, *Tales from a Northwest Naturalist*, is packed with stories about his adventures with animals.

This particular day, though, Jim and I planned to fly—just for fun.

I arrived at Jim's house at breakfast time and found him in the kitchen, mixing up his special hotcake batter. Gosh I loved those hotcakes! Jim had just finished pouring the first cakes on the griddle when, much to our amazement, a large, on-the-mend barn owl swooped out of a bedroom, plopped feet-first in the batter, and landed on the sizzling griddle. Frightened, the owl bolted into the air and ricocheted around the house, spewing hotcake batter along the way.

Finally, when the owl flew in for a landing in the living room, Jim snagged his patient and returned it to the rehabilitation room.

By this time, of course, Jim and I were laughing so hard we could hardly speak. But Jim went right back to stirring and pouring his hotcake batter.

Cousteau

Back in the day, being in the business of exotic animals meant comingling in a small world of experts, so it wasn't surprising that I got to know Jacques Cousteau, the famous oceanographer, and his sons. We held a lot of meetings, conferred, and talked about animals in general, usually over coffee, here and in California. I had done quite a bit of work on sea lions, seals, and whales, and he was very knowledgeable about them as well.

Cousteau was intelligent and a hard worker . . . and I appreciated that.

Morgan and Ditty

I've already told you about my good friend Morgan Berry, but another treasured friend also deserves mention. Dr. Marlowe Dittebrandt, who I believe was Portland's first female physician, worked at the zoo as a pathologist in the late 1960s. "Ditty," as we called her, became good friends with both Morgan and me—first with me at the zoo, then as Morgan's assistant at nearby Elephant Mountain.

One year, probably in the late '70s, Ditty agreed to accompany Morgan on a winter trip to St. Paul, Minnesota. They went to get an elephant named Mi Thai, then living in that city's zoo.

The story picks up in Minnesota, after Morgan and Ditty get Mi Thai and leave town, headed straight into a snowstorm. Ditty, who lobbied hard against driving in the snowstorm, writes in her travelogue:

Morgan's eyes are fixed straight ahead. He is determined to reach Albert Lea [south of St. Paul on I-35] no matter what. The "no matter what" comes quickly: We drive into a white out. I plead with him to slow down or pull over to the roadside. No luck.

An apparition appears dead ahead. Morg hits the brakes and we hit ice. Around and around we go—first we are ahead of the trailer and then the trailer passes us. The spin ceases when the pickup bangs into a flatbed trailer stalled crosswise in the road. Mi Thai doesn't tip over, but the tack compartment is smashed open. Neither of us appears to be hurt and the windows are not broken. Morgan sits staring straight ahead, as if in shock. I open the truck door to check the trailer damage. Flames are licking up the walls of the tack compartment, and one of the bales of hay in the truck has started to burn. "The trailer's on fire!" I scream.

You've probably noticed that Morgan has (don't we all?) some irritating characteristics. However, when an emergency faces him, he is fearless and instantly responsive. My scream of fire [has] scarcely faded before Morgan is out of the truck.

I follow with the fire extinguisher, which, of course, proved absolutely useless in blizzard wind. Morgan grabs the burning hay and throws it from the truck. The wind rolls it out across the flat fields. The windchill factor, which we learned was minus 40, is too much for me. I can scarcely get a breath. I retreat to the pickup and jump inside. Not Morg. I watch him grab the flaming coats and shirts and throw them to the winds. He empties the compartment of flaming heaters and splintered wood. He stomps the fire from a blanket and then uses it to smother the remaining flames. All of this, he accomplishes in no more than two minutes.

Morgan and Ditty made it home safe with Mi Thai—though Morgan's determined attitude again sent them spinning on icy roads in New Mexico (they traveled a southern route, hoping to avoid more bad weather). After that harrowing adventure, Ditty never traveled farther than Seattle with her good friend Morg.

Ditty died in the 1990s, years after Morgan. She never married. Switching careers in midlife takes guts, and Ditty had plenty. In the male-dominated 1960s, she was one of us.

Ditty was a woman ahead of her time.

Pies and Jam

Uncharacteristically for an outdoor guy like me, I got into making jams and pies in my eighties—I guess because it was something new to try. Patricia's grandmother was a test cook for Betty Crocker, which made Patricia a good teacher. After making my first few jars, my

Later in life I learned how to make jams and jellies, and entered them in the county and state fairs. I must be doing something right, because I've sure won a lot of ribbons. (Photo by Patricia Maberry)

granddaughter and my daughter-in-law encouraged me to enter my jams in the Washington County Fair. I did just that, and I got a bunch of ribbons, so I kept on going and competed in the Oregon State Fair. Every year since then I have received a fistful of ribbons at various fairs.

I make strawberry, raspberry, apricot, pineapple, boysenberry, and plum jams. I also bake pies, cakes, biscuits, and bread. If I'm going to do something, I always get in and learn as much about it as I possibly can. I always want to see if I can do it better.

Mucking in the Moonlight

One night Patricia and I received a call from our good friends Howie and Donna Renner, telling us that one of their spring calves was due and that the mother was acting particularly nervous. Howie, an industrial engineer, owned what's now known as a "gentleman's farm" in Wilsonville, Oregon, meaning he worked a nine-to-five job and managed a small, personal farm. Howie remains one of my closest friends; the two of us often go on field trips to places like the Seaside Aquarium and the Evergreen Aviation & Space Museum in McMinnville, Oregon, home of Howard Hughes's famous *Spruce Goose*.

That night, though, my mind was on the yearling mother, not field trips. The summer before, a neighbor's big bull had jumped the fence and come a courtin'. Sometimes such a mating results in a huge calf and subsequent problems during birth.

Patricia and I arrived at the Renner's farm around 11 PM and soon spotted the bawling mama cow across the field, standing in the pouring rain. The cow's frightened eyes reflected in my flashlight beam, and I could see two hooves bouncing from her backside. Patricia and Donna trailed Howie and me, trying to carry my fifty-pound bag and other necessary equipment. Much colorful language poured from my wife's mouth as she and Donna slogged their way through thick mud and water! Howie and I, on the other hand, held only the lightweight rope we hoped to wrap around the cow's neck so she could be tethered to a tree during delivery.

Any and all plans went up in the air when the terrified animal trotted off down a small hill to a creek below. When I finally got down in that creek and tried my best to rope her, off she went again. This time the rest of Howie's cattle herd trailed the wailing mama.

Now I was really starting to worry. We were all pretty tired, and the odds for a successful delivery were waning.

But we rallied, picked up our gear again, and trudged across the meadow toward the expectant mother. That's when the moonlight suddenly split the clouds and beamed down upon us. It was a small moment of shining hope . . . that quickly dimmed as the mother dashed into a field thick with holly bushes and trees.

My comfortable bed at home was sounding better and better.

But luck prevailed! The poor cow was so scared and so weary that she finally let me tie her to one of the holly trees. I dispatched Howie and Donna for water and blankets, expecting the worst for the delivery; I didn't know if the baby calf was even still alive.

With one last-ditch effort, I roped the baby's feet, and Patricia and I started pulling on the count—ONE, TWO, THREE, *PULL!!!* This is when Patricia began using her colorful words again to encourage that cow to, by golly, *push*. It seemed like forever that we were counting and pulling, when finally, after a particularly big yank, we felt a *whoosh* and heard a *plop*.

Then, silence.

We feared bad news, but then . . . a hearty *baaaaaa* reverberated the air. The calf lived! We laughed and we cried at the sound of new life. Happily, I tended the mother to make sure all was well. With peace restored, we left mother and baby alone to bond, nuzzle, and nurse.

The four of us "obstetricians" first commiserated about our midnight adventure but ultimately congratulated one another on a job well done. Mucking in the moonlight in the middle of the night really wasn't so bad after all.

Like so many other moments in this crazy life I've led, it was all rather *magical*.

"I often wish I had more time; there's still so much to learn. My life has been blessed with many adventures and grand experiences. Packy, of course, tops the list.

(Reprinted with permission from The Oregonian*)*

“*When you are around elephants you can't help but fall in love with them.*

—Doc Maberry

PACKY'S FAMILY TREE

BELLE
Born: 1952 (Thailand)
Sold: 1962 to Oregon Zoo

THONGLAW
Born: 1947 (Cambodia)
Sold: 1962 to Oregon Zoo

ROSY
Born: 1951 (Thailand)
Sold: 1953 to Oregon Zoo

PACKY
Born: 4-14-1962 (Oregon Zoo)
Lives in: Oregon Zoo

ME-TU
Born: 1962 (Oregon Zoo)
Lived in: Oregon and Los Angeles Zoos

SUNG-SURIN (SHINE)
Born: 12-26-1982 (Oregon Zoo)
Lives in: Oregon Zoo

RAMA
Born: 4-1-1983 (Oregon Zoo)
Lives in: Oregon Zoo

HIGHLIGHTS

1796: First elephant arrives in United States on trade ship *America*.

1880: Hebe the circus elephant delivers Little Columbia, believed to be America's first infant pachyderm.

1882: Second captive elephant is born, named for its birthplace, Bridgeport (Connecticut).

1888: Portland's first zoo opens with two caged bears at City Park, later renamed Washington Park.

1910: Circus elephants Alice and Snyder breed first of four ill-fated captive babies born between 1912 and 1918.

1917: Matthew Bartram Maberry is born in Seattle, Washington, on November 22.

1918: Birth of Prince Utah, last captive elephant born in Western Hemisphere—until Packy. He dies nearly a year later.

1925: Portland's zoo moves to current site of Japanese Garden in Washington Park.

1947: Doc Maberry graduates from veterinary school at Washington State University; new Zoo Director Jack Marks pushes Portland for better zoo.

1952: Morgan Berry buys Packy's mother, Belle, in Thailand.

1953: Austin Flagel donates zoo's first elephant, Rosy.

1954: Portland voters pass $3.85 million ballot measure to fund new zoo.

1956: Orville Hosmer donates zoo's second elephant, Tuy Hoa.

1958: Doc becomes zoo's first full-time vet; Jack Marks brings back penguins from Antarctica.

1959: Portland Zoological Gardens opens in July at current Oregon Zoo site.

1960: On July 17 Belle and Thonglaw mate at Seattle's Woodland Park Zoo.

* * *

1962

January

Doc announces to media that Belle, Rosy, and Tuy Hoa are pregnant; also suspects Pet is expecting.

January–mid-April

Media camps in elephant barn, awaiting baby elephant; "Expectant Elephant" article appears in Newsweek magazine.

April 14

At 5:58 AM, Belle delivers Packy.

May

Article on Packy's birth appears in *LIFE* magazine.

June

Morgan Berry sells Belle and Packy to Portland, and includes Thonglaw and Pet in deal.

August

Packy cuts first tooth; in lifetime, he'll chew through six sets of teeth.

October 2

Rosy delivers Me-Tu, so named because her birth is less publicized than Packy's.

By year's end

International Zoo Yearbook publishes Doc's paper "Breeding Indian Elephants."

* * *

1963: In September, Tuy Hoa gives birth to Hanako; Pet delivers Dino.

1965: Morgan Berry creates animal menagerie on eighty-acre Elephant Mountain near Woodland, Washington.

1968: Doc presents paper "Delivering Baby Elephants" to American Association of Zoo Veterinarians.

1971: Jack Marks retires; Portland Zoological Society assumes zoo management, hires new director. *International Zoo Yearbook* publishes Doc's paper "Mummified Foetuses in a Bactrian Camel at Portland Zoo."

1973: Doc spends year in bed battling rare herpes virus; retires from zoo.

1975: Doc and Patricia marry, form Wildlife Enterprises.

1976: Metropolitan Service District (Metro) now runs zoo, renames it Washington Park Zoo; Doc goes to China, hoping to purchase pandas.

1978: Oregon Historical Society tapes oral history interview with Doc.

1980: Doc becomes vet at Seaside Aquarium, a post he still holds.

1986: Doc shares animal tips on *Prime Timers* TV show.

1998: Metro renames zoo: Oregon Zoo.

2000: Writer Shana Alexander dedicates her book *The Astonishing Elephant* to Doc.

2002: Doc returns to zoo to celebrate Packy's fortieth birthday.

2007: Doc appears on Oregon Public Broadcasting's *Oregon Field Guide*.

2008: Doc receives awards for lifetime work; also coaches zoo staff before Samudra's birth.

2009: Oregon Zoo celebrates fifty-year anniversary.